Vibrations of Thick Cylindrical Structures

Hamid R. Hamidzadeh · Reza N. Jazar

Vibrations of Thick Cylindrical Structures

Springer

Hamid R. Hamidzadeh
Department of Mechanical Engineering
Tennessee State University
3500 John A. Merritt Blvd.
Nashville TN 37209-1561
USA
hhamidzadeh@tnstate.edu

Reza N. Jazar
Department of Mechanical
and Automotive Engineering
School of Aerospace,
Mechanical and Manufacturing Engineering
RMIT University
Melbourne, VIC 3083
Australia
reza.jazar@rmit.edu.au

ISBN 978-0-387-75590-8 e-ISBN 978-0-387-75591-5
DOI 10.1007/978-0-387-75591-5
Springer New York Dordrecht Heidelberg London

Library of Congress Control Number: 2009937454

Printed on acid-free paper

Springer is part of Springer Science+Business Media (www.springer.com)

Contents

List of Figures

Dedicated to our wives *Azar* and *Mojgan*.

Happiness is thinking by yourself,
not by someone or something else,
not for someone or something else.

Preface

The authors have endeavored to develop a book that is the result of over three decades of academic careers, conducting research on various aspects of Vibration Engineering, and dissemination of their research results. The book introduces the reader to the analytical techniques for vibration analysis and damping of thick cylindrical structures where linear elastic shell theories cannot offer realistic solutions in this important engineering problems. This book presents necessary theoretical background for design analysis. Furthermore, it offers the engineering information and quantitative data needed for design analysis, applications, and construction of these structures. The book, in its entirety, constitutes an extensive guide for the reader. It also provides a systematic solution for the modal analysis of thick cylindrical structures and panels, and extends the solution to vibration of layered cylindrical structures.

The book is intended to lay the foundation for understanding mathematical modeling, vibration analysis, damping, and the design of thick elastic and visco-elastic single-layer and multi-layer cylindrical structures in a complete and succinct manner. Throughout the book, an attempt has been made to provide a conceptual framework that includes exposure to the required background in mathematics, fundamentals of the theory of elasticity, modal analysis, and constrained-layer damping treatment, as well as Bessel differential equations and their solutions. The knowledge of the presented topics will enable the reader to pursue further advances in the field.

Level of the Book

The primary audience of this book is the graduate students in mechanical engineering, engineering mechanics, civil engineering, aerospace engineering, ocean engineering, mathematics, and science disciplines. In particular, it is geared toward students interested in enhancing their knowledge by taking the second graduate course in the areas of vibration of continuous systems, structural dynamics, and application of passive damping treatments. The presented topics have been prepared to serve as an aid to designers of cylindrical structures. It can also be utilized as a guide to practicing engineers who are responsible for setting design specifications and ensuring their fulfillment. The covered topics are also of interest to research engineers who are seeking to expand their expertise in these areas.

Organization of the Book

The book is presented in seven chapters: introduction, fundamentals of elasticity, vibration analysis for single-layer cylinders, modal analysis for single-layer cylinders, vibration of multi-layer thick cylinders, constrained-layer damping for cylindrical structures, and vibration of thick cylindrical panels. Furthermore, it offers helpful and significant tabulated results, which can be used as design guidelines for these structures.

To make effective use of the presented topics, the following procedure is suggested. A working understanding of certain theoretical concepts and method can be achieved by referring to chapters 1 through 3. To become acquainted with the state of the art in this particular field and learn about the historical background on this topic, the reader should begin with chapter 1, which contains an extensive list of key references with brief discussions on their methodologies, required assumptions, and their achievements. Chapter 2 reviews the succinct fundamental theoretical background and concepts needed from the theory of elasto-dynamics, which will enable the reader to follow the derivation of the required governing equations and their solutions in chapter 3. Chapter 4 is intended to present numerical results for the natural frequencies and mode shapes of cylindrical structures with a variety of geometries. Chapters 5 and 6 present analytical modeling, and informative case studies for vibration analysis of the multilayer cylindrical structures and the constrained layer damping treatment of these structures. Chapter 7 is designed to provide the essential modal analysis for long, thick, cylindrical panels based on the plane strain assumption.

Method of Presentation

The scope of each chapter is clearly outlined and the governing equations are derived with an adequate explanation of the procedures. The covered topics are logically and completely presented without unnecessary overemphasis. The topics are presented in a book form rather than in the style of a handbook. Tables, charts, equations, and references are used in abundance. Proofs and derivations are often emphasized, and the physical model and final results are accompanied with illustrations and interpretations. Specific information that is required in carrying out the design analysis in detail has been stressed.

Prerequisites

This book is primarily intended for graduate students, so the assumption is that the readers are familiar with the fundamentals of vibration engineering and differential equations, as well as a basic knowledge of linear algebra and numerical methods. The presented topics are given in a way to establish a conceptual framework that enables the reader to pursue further advances in the field. Although the governing equations will be derived with adequate explanations of the procedures, it is assumed that the readers have a working knowledge of mechanical vibration and the theory of elasticity.

Unit System

To maintain generality, computed results and the required parameters are provided in non-dimensional forms. Nevertheless, the system of units adopted for case studies is, unless otherwise stated, the British Gravitational system of units (BG). The units of degree (deg) or radian (rad) are utilized for variables representing angular quantities.

Acknowledgements

We wish to express our thanks to our colleagues who have assisted in the development of this book. We are indebted to our friends and colleagues Hurang Hu, Albert C. J. Luo, and Annie Tangpong for their constructive comments and suggestions.

1

Introduction

Due to strong potential applications and more demanding requirements imposed upon applications of cylindrical structures, there has been increasing research and development activities during recent years in the field of vibration analysis and damping treatment for this type of structure. An important step in vibration analysis of cylindrical structures is the evaluation of their vibration modal characteristics, such as natural frequencies, mode shapes, and modal loss-factors. This modal information plays a key role in the design and vibration suppression of these structures when subjected to dynamic excitations. Most reported studies on the dynamic response of cylindrical structures have been restricted to the application of shell theories. These theories are based on a number of simplifying assumptions, the most important of which is that the considered shell must be relatively thin to assume constant stresses within the cylinder. Therefore, due to this limitation, shell theories are inadequate to accurately describe many modes of vibrations which occur in thick cylindrical structures. The primary scope of this chapter is to address a brief historical background on the key reported investigations on this topic.

It should be noted that there is extensive literature available on vibration analysis and damping treatments of thin cylindrical shells and panels. However, due to the main scope of this book, this chapter is more concerned with reporting the research results for thick cylindrical structures. Nevertheless, some major reported developments addressing the vibration analysis and damping treatments for thin cylindrical structures are also presented.

1.1 Vibrations

The phenomenon of vibration was used by ancient peoples in attempts to produce music. One of the first widely recognized mathematical studies of vibration was Galileo's study of pendulums and the transfer of energy between two bodies tuned to the same natural frequency (Timoshenko, 1953). Galileo continued his research into the vibration of strings. Along with Hooke, Galileo showed the relationship between sound and the frequency of mechanical vibration (Timoshenko, 1953). Many mathematicians

H.R. Hamidzadeh, R.N. Jazar, *Vibrations of Thick Cylindrical Structures*
DOI 10.1007/978-0-387-75591-5_1, © Springer Science+Business Media, LLC 2010

contributed to the study of vibrations. Some of the prominent mathematicians include Taylor, Bernoulli, D'Alembert, Euler, Lagrange, and Fourier (Rao, 1990). Wallis and Sauveur independently observed the mode shape and node formation phenomenon in the vibration of strings (Cannon and Dostrovsky 1981). Bernoulli was credited with the principle of linear superposition of harmonics (Cannon and Dostrovsky 1981).

In the seventeenth century, Hooke presented his law of elasticity. Euler and Bernoulli used Hooke's law to produce the governing differential equations for the lateral vibration of prismatic bars, and solved the equations for small displacements, (Rao, 1990). Coulomb expanded upon their research by investigating the torsional vibration of wire and presenting the results. One of Lagrange's many contributions was in correcting the work of Sophie Germain in 1809. This research mathematically described the theory of plate vibration. Lagrange's contribution was later corrected by Kirchhoff in 1850 (Bucciarelli and Dworsky, 1980).

In 1877, Lord Rayleigh published his book on the theory of sound. He later produced a method for finding natural frequencies by the principle of conservation of energy (Lindsay, 1970). Rayleigh's work is still significant today. Vibration research is also described as the study of wave propagation. One of the most historically important studies of wave propagation was produced by Horace Lamb (1904). Lamb studied the propagation of vibrations over the surface of a "semi-infinite" isotropic elastic solid, more conventionally known as an elastic half space, subjected to an impulse load at an arbitrary point on the surface.

Modern contributors to the theory of vibration are Timoshenko, Stodola, and Mindlin. Stodola added to the topic of turbine blade vibration. Timoshenko and Mindlin improved theories about the vibration of beams and plates. Most modern textbooks that introduce mechanical vibration contain many direct references to Timoshenko's work. For the interested reader, textbooks written by Timoshenko (1953) contain many of the same theories that are currently in use.

1.2 Vibrations of Cylindrical Structures

Cylindrical structures are widely used in the construction of structures from the fuselage of a jet airplane or missile to pressure and piping vessels. A greater understanding of these structural components is desired to prevent failure and to optimize their design. This book is intended to present vibration analysis and the modal characteristics of these structures. The identification of the resonant frequency of these structures and the effect of various levels of material damping on the response are intended to help in the design of these systems. Thus, the effect of different elastic and visco-elastic materials on the vibration of a typical thick cylinder is considered. This book begins with descriptions of the technique used in the study of the forced vibration of visco-elastic cylinders. It then concerns with the study of the relation between the material loss-factor and the modal damping loss-factors associated with different mode shapes.

1.3 Vibrations of Thin Shell Cylinders

Due to the complexity of using the theory of elasticity for all cylindrical structures, the theory of shells was developed to address dynamic analysis of thin cylindrical structures. Extensive research is reported for different aspects of vibration, stability and control of the cylindrical structures. In the majority of the reported research, it was common to use shell theories. However, the theory of shells makes many approximations. It should be noted that the shell considered must be relatively thin to justify constant stresses within the thickness of the shell.

Several techniques have been used to study the vibrations of hollow cylinders. Initially, the simplest technique, the shell theory, was used. Rayleigh (1894) determined the natural frequencies of thin cylinders with free ends while considering axial vibration. Love (1927) studied the flexural vibration of cylinders. Flugge (1934), by a similar approach, succeeded in obtaining a frequency equation for a cylinder with freely supported ends. The three roots of his equation defined three natural frequencies for any given nodal pattern. The problem was further investigated by Arnold and Warburton (1949). They attempted to explain certain frequency phenomena that were observed in experiments on thin cylinders. For a cylinder with freely supported ends, they derived the frequency equations and were able to verify the experimental results with considerable accuracy. They found that for short cylinders with very thin walls, the natural frequencies may actually decrease as the number of circumferential nodes increases. This was shown theoretically to be due to the proportion of strain energy contributed respectively by bending and stretching.

Arnold and Warburton (1953) considered the flexural vibrations of thin elastic cylinders. In this type of vibration, many forms of nodal patterns may exist due to the combination of circumferential and axial nodes. Theoretical expressions were developed for the natural frequencies of cylinders with freely-supported and fixed ends. They reported an acceptable agreement between the theoretical and the experimental natural frequencies.

Forsberg (1964) implemented Flugge's equations of motion to determine natural frequencies for uniform thin shells, regardless of the end conditions. His method assumed a natural frequency and sought a length of cylinder, this was different than all other methods previously used. Warburton (1965), using Forsberg's method, investigated thin shells with clamped and free ends. Warburton's concluded that for long wavelengths and large numbers of circumferential waves the effect of the end conditions was negligible. Therefore, it was possible to treat the ends as simply supported for frequency calculations. An account of the various thin shell methods can be found in the introductory chapter of Thin-Shell Structures: Theory, Experiment, and Design, edited by Fung and Sechier (1972). Extensive discussions on the applications of thin shell theories are provided in a NASA Technical Report, by Leissa (1973), and also in Pietraszkiewiz and Szymcvak (2006), and Amabili (2008). Needless to say, the shell theories are inadequate to accurately describe many modes of vibration in shells, as well as thick cylinders. Therefore, due to the limitations associated with the shell theory, the elasto-dynamic theory should be employed.

1.4 Vibrations of Rods

As researchers gained interest in the dynamics of cylinders, the vibration analysis of solid shafts was considered. Pochhammer (1876) was the first to study the vibrations of solid cylinders. Chree (1889) studied infinitely long circular rods with traction-free circumferential boundaries. Both investigators utilized the theory of elasticity in their research. Using the theory of elasticity, McNiven and Perry (1962) discussed the coupling of longitudinal axial shear and radial modes in an infinitely long rod. McNiven et al. (1966) proposed a "three-modes theory" for rods that compared well with elasto-dynamic theory. In this study, the first three modes of vibration are assumed to be dominant, while the other modes are assumed to contribute little to the overall response.

Rumerman and Raynor (1971) studied the radial and axial modes of an infinitely long cylinder using the Rayleigh-Ritz method. Since structural cylinders are not of infinite length, the next logical step was to investigate rods of finite length rods. Hutchinson (1972) studied axisymmetric free vibrations of finite length rods using a series solution to the three-dimensional theory of elasticity. McMahon (1964) published his experimental results for natural frequencies of solid cylinders with free boundaries. These important experimental results are not adequate for determination of modes and geometric or elastic parameters. Rasband (1975) extended Hutchinson's earlier work to study non-axisymmetric free vibrations in a solid cylinder of finite length. Hutchinson (1980) then extended his work to include non-axisymmetric vibrations of a solid rod and compared his results to those obtained experimentally by McMahon (1964). Lusher and Hardy (1988) studied axisymmetric free vibration of a transversely isotropic finite cylindrical rod.

1.5 Vibration of Thick Cylindrical Structures

The shell theory may be applied to many thin cylindrical structures, while thicker cylinders can be accurately studied only by using the three-dimensional theory of elasticity. It should be noted that the theory of elasticity can be applied to thick or thin cylindrical structures. Among the earlier thorough studies of infinitely long traction-free hollow cylinders, using the theory of three dimensional elasticity, one can mention the independent works reported by Greenspon (1957) and Gazis (1959).

Gazis (1959) studied the free harmonic wave propagation along a hollow circular cylinder of infinite extent, within the framework of the linear theory of elasticity. He derived a characteristic equation appropriate to the circular hollow cylinder by using the Helntholtz potential for arbitrary values of the physical parameters involved. Armenakas (1967) derived the frequency equation for harmonic waves, with an arbitrary number of circumferential nodes, traveling in traction free composite, circular cylindrical structures. Armenakas et al. (1969) considered the transmission of elastic energy by means of elastic waves, and formulated the eigenvalue problem for stress-free cylinders for different circumferential wave numbers. They also presented tables of non-dimensional natural frequencies for a wide range of geometrical possibilities. Their analysis was based on the linear three-dimensional theory of elasticity. In a separate article, Armenakas (1971) provided numerical results for evaluating the effect of changes of the shell parameters on the frequency and shape of the first few modes of vibrations. The composite shells, used in Armenakas et al. (1969), consisted of infinitely long cylinders with two concentric circular shells bounded at the interfaces.

Hutchinson (1979) studied the axisymmetric flexural vibrations of thick free circular plates by using the general three-dimensional linear elasticity to find accurate natural frequencies and mode shapes for the flexural vibrations of free circular plates. He included shear and rotary inertia effects, compared them with the accurate series solution used by Pickett (1944), and found that the approximate solution yielded frequencies with sufficient accuracy for most engineering applications. Hutchinson (1980) also applied the series solution method to find the natural frequencies of solid vibrating cylindrical rods. Hutchinson and Ei-Azhari (1986) extended Hutchinson's work in solid cylinders to include free hollow cylinders with finite length.

Other approaches for the vibration analysis of hollow cylindrical structures have been proposed. McNiven et al. (1966) used a "three-mode theory" to obtain good agreement with the exact solution based on the three-dimensional theory of elasticity. The three-mode theory is based on the assumption that the first three (lowest) modes have a high influence on each other and the other modes are less significant. Gladwell and Vijay (1975) used a finite element approach to the three-dimensional vibration of finite length circular cylinders with traction-free surfaces and free-end conditions. Svardah (1984) treated wave propagation in semi-infinite, elastic cylinders with traction-free lateral surfaces initially at rest and excited by transient end loading. Markus (1988) studied the effect of visco-elastic material on the vibration of cylindrical structures using the shell theory. Singal and Williams (1988) developed approximate solutions for natural frequencies of thick cylinders and rings, and verified their results by conducting experiments.

Stephen and Wang (1992) employed the Papleovitch-Nueber solution for the elasto-static displacement equations of equilibrium, and derived both axial and non-axisymmetric solutions. They considered zero traction on the surface generators of the infinitely long elastic cylinder and formulated an eigen equation for free vibration. They computed smaller roots of the eigen equation for the circumferential harmonic numbers of $n = 0$, 1, 2 and 3 for different wall thickness ratios. A formal mathematical review of elasto-dynamic problems, such as the one presented here, is presented by Mal and Singh (1991).

The free vibration solution to hollow cylindrical dynamics prompted the next step of research which was to determine the forced response of the cylinder. Other interesting work in the area of forced vibrations can be found in research work reported by Braga et al. (1990). Braga's research consisted of the study of wave propagation in a fluid-loaded laminated cylinder The effects of the laminates and cylinder curvature on the dispersion curves of the subsonic interface waves were given particular attention. In this study, the three-dimensional elastic theory was used for accurate analysis of acoustical waves. This investigation is of interest due to the similar procedure for choosing appropriate Bessel function values for the solution to the elasticity problem. Several forced vibration solutions have also been reported using the shell theory. When using this theory, many dynamic effects have been neglected in order to obtain a closed form solution. An example of this is the impulse loading of orthotropic cylinders by Christoforou and Swanson (1990) using the shell theory. Many forced vibration problems have not yet been solved using the linear elastic theory.

Hamidzadeh et al. (1991) studied the forced vibrations of hollow cylinders. They employed the three-dimensional elastic theory and developed the governing equations of motion in terms of volumetric strain and elastic rotations. They then produced a system of linear non-homogeneous equations and solved for the stresses and displacements for a wide range of excitation frequencies. Using a frequency sweeping procedure, they calculated resonant frequencies and compared their results to those of Gazis (1959) and found dissimilar results for short cylinders and fundamental resonant modes.

Hamidzadeh et al. (1991), and Hamidzadeh and Moxey (2004) also studied the harmonic responses of visco-elastic thick circular cylinders of infinite extent. The cylinders were subjected to harmonic radial and tangential boundary stresses. The two-dimensional theory of elasticity and a complex elastic modulus were used to provide a solution for stresses and displacements at any point in the cylinder. Hamidzadeh and Chandler (1991) extended their research work to develop a two-dimensional visco-elastic theory involving multi-layered sandwich cylinders. The solution to the three-layer cylinders was developed by utilizing the stresses and displacements for the single cylinder and by complying with the compatibility requirements at each interface. Their computed resonant frequency for a three-layer elastic cylinder with the same properties compared satisfactorily with those computed for a single-layer cylinder having identical geometry.

1.6 Vibrations of Damped Cylindrical Structures

The main objective in using visco-elastic material is to dissipate energy and reduce the amplitude of the vibration. The visco-elastic material can be mathematically modeled using complex stiffness and moduli. This is due to the phase difference between stress and strain. Visco-elastic materials are very resilient and are widely used for damping treatments. The use of such material reduces the amplitude of stresses and displacements for a longer working life. This section contains an overview of the past research works on the effect of material damping on vibration characteristics of thick cylindrical structures. A notable work was conducted by Parfitt and

Lambeth, (1962) who studied the damping and transmission of waves in visco-elastic material structures. Warburton (1965) showed the response for shells under loading, for different material loss-factors.

While most research for determination of the modal analysis area is focused on finding the actual resonant frequencies of the cylindrical structures, few have sought to find the actual effect of the differing geometric and material properties on the vibration of the cylinders. Lazan (1968) uses the definition of rheology to include elastic deformation, plastic creep flow, creep relaxation, and damping. Flugge (1967) and Snowdon (1968) have shown the process of considering "rubber-like materials" as having a complex elastic modulus by considering the relationship of stress to strain with linear partial derivatives.

Ewins (1986) has given a mathematical method for determining the modal loss-factor of structures based on circle-fitting the response data in the vicinity of the resonance. Yeh and Yang (1987) have described a time domain technique for the modal identification of a system. Mead and Markus (1970) determined modal frequencies and loss-factors for an incasters damping sandwich beam. Crandall (1991) has described damping models in the time domain for the ideal viscous damper, the ideal hysteretic damper, and the band-limited hysteretic damper. Pang et al. (1992) and Bergman (1993) considered the effect of visco-elastic materials on vibration control of actuators and hybrid control of Euler-Bernoulli beams.

Hamidzadeh and Chandler (1991) applied the general three-dimensional theory of elasticity to study the harmonic forced vibration of thick visco-elastic hollow cylinders. Resonant frequencies of the breathing and first three lobar modes for cylinders were computed. Their results compared well with those of Armenakas et al (1969).

1.7 Vibration of Multilayered Cylindrical Structures

Composite multilayer cylinders have widespread use in engineering. Pipes, aircraft, submarines, missiles, rockets, and power transmission shafts are typical cylindrical structures. When the frequency of an acting force on a cylinder coincides with the natural frequency of the structure, the phenomenon of resonance occurs. Resonance is associated with large displacements and stresses and may cause failure in the cylinder. Therefore, determining the natural frequency is vital and has a direct impact on the design of composite cylindrical structures.

Many researchers have investigated the vibration of cylinders using different analytical approaches, such as: the thin shell theory, the curved beam theory, and the theory of elasticity. Most studies on the vibrations of constrained layer damped cylinders are based on the assumption of thin shells. The assumptions used in the shell theory for cylindrical structures may not always be valid. In fact, a significant number of cylindrical structures are considered to be thick. In spite of this, few studies have been directed to the analysis of the free vibrations of thick cylindrical structures.

Crampin (1970) studied desperation of surface waves in multi-layered and isotropic media. Bhaskar and Varadan (1990) developed a solution for analysis of pinched laminated cylinders. Rattanawanqcharoen and Shah (1990) studied the wave propagation in a laminated isotropic cylinder. The dispersion relation for a multi-layered elastic cylinder was evaluated by a propagator matrix and nondimensional natural frequencies for a two-layered cylinder with a special circumferential wave number, $n = 1$, and a range of axial wave numbers were calculated. Hawkes and Soldatos (1992) studied the vibrations of multi-layered laminated cylinders. They achieved the solution using a method of successive approximations. The exact governing equations of the cylinders were replaced by a set of simpler approximate equations that were solved analytically. They compared their results with those of Armenakas et al. (1969).

Huang and Dong (1984) utilized a stiffness method to consider wave propagation in laminated anisotropic cylinders with an arbitrary number of lamina. Braga and co-workers (1990) studied harmonic wave propagation in a fluid-loaded composite circular cylinder, where the cylinder was composed of isotropic elastic layers. In their computation, they developed an analytical method to determine the impedance of laminated elastic cylinders.

Braga et al. (1990) considered the wave propagation in fluid-loaded laminated cylindrical shells. Hamidzadeh and Sawaya (1995) studied the free vibration of multilayer visco-elastic cylinders. To introduce visco-elastic damping in the composite cylinder, their analysis allowed complex shear moduli for each layer. The problem was expressed in the form of matrix equations relating the displacements of a point in the medium to the boundary stresses. Their analytical approach for a single frequency independent visco-elastic layer was developed to obtain the displacements and stresses in terms of boundary stresses for any given circumferential and axial wave number. This formulation was used to develop a relation between displacements and stresses at the boundaries of each layer and consequently to obtain a propagator matrix relating displacement and stresses of the inner boundaries to the outer ones. Invoking the zero boundary stresses, the analysis led to a frequency equation for the free vibration of the multilayer visco-elastic cylinder. The presented analytical procedure was also applicable to multilayer cylinders of arbitrary laminated configuration and material damping for each layer. Zhou and Yang (1995 and 1996) developed a distributed transfer function method for vibration analysis of thick laminated composite cylinders. Qatu (1999) employed Dong's solution (1968) and provided analysis for laminated composite shells. He considered the free vibration analysis and evaluated frequencies for orthotropic and isotropic cylindrical shells. Qing et al. (2006) presented a semi-analytical solution along with a finite element method for free vibration of thick double layer cylinder.

1.8 Vibrations of Constrained Layered Cylindrical Structures

Damping is a significant dynamic parameter for vibration, sound control, dynamic stability, positioning accuracy, fatigue endurance, and impact resistance. Many current applications of the circular structure, (aerospace industry, power generation or transmission, heat exchangers, tubes, nuclear reactor components, smoke stacks, and undersea technology) require high dynamic performance. Therefore, candidate sources of passive damping should be compatible with structural configuration. To control large displacement and stresses, damping treatment using visco-elastic material with complex shear modulus is employed. In the constrained layer damping treatment, a visco-elastic material is sandwiched between two elastic layers. The visco-elastic layer dissipates much of the energy within the visco-elastic cylinder in the form of heat during shear de-formation when subjected to harmonic vibration. In this configuration, the core layer allows shear displacement and heat dissipation, while the elastic layers give the cylindrical structure rigidity and stability.

The appropriate use of the visco-elastic damping material is to sandwich it between two elastic materials. The elastic material will give the structure stiffness, while the damping material will dissipate much of the energy. With the continuing evolution of the composite structures and simultaneous requirements for higher performance and lower operational costs, more and more composites are targeted toward structural applications involving combined dynamic, mechanical, thermal, and hydraulic loading. Composite structures are principally preferred in such applications because of their advanced elastic properties for individual design requirements. They also have the potential for incorporating significant passive damping into the candidate structure.

The potential damping sources satisfying the previous requirements are the constrained layer damping approach. Constrained damping in isotropic metallic structures has been applied and investigated. The inherent damping capacity of composite structures also seems promising. Although the damping of composite structures is not very high, it is significantly higher than most of the common metallic structures. Moreover, visco-elastic damping layers are the material of preference in many cases, as they readily provide high specific stiffness and strength.

Constrained-layer cylindrical structures with visco-elastic damping layers will offer the advantages of high damping and good mechanical properties. In addition, the visco-elastic damping concept is highly compatible with the laminated configuration of composite structures. A visco-elastic layer may produce higher and also tailorable damping. However, other critical mechanical properties, such as stiffness and strength, are expected to be reduced. It is desirable to develop integrated composite mechanics that will efficiently predict the damping of composite structures with a visco-elastic damping layer, and other critical mechanical properties and dynamic characteristics. An analytical work including the damping analysis of three-layered cylinders with constrained inter-laminar layer of visco-elastic material has been reported. The effects of different geometries and damping layer thicknesses are also investigated.

Research on the constrained layer damping treatment for beams was conducted by Agbasiere and Grootenhuis (1968). In this study, they transformed the equations of motion for multilayered beams with visco-elastic cores into non-dimensional simultaneous equations and used finite difference methods for their solution. It was shown that having a damping layer in shear strain in between two rigid layers was effective. Grootenhuis (1970) demonstrated the variation of loss-factor, frequency parameter and tip amplitude ratio for a three-layered symmetric cantilever beam. Nakra and Grootenhuis (1972) presented the variation of loss-factor versus shear parameter for a different number of layers in beams and concluded that dual visco-elastic cores are superior to either no core or a single core. Nakra and Grootenhuis (1974) presented the effect of extensional properties on the loss-factor of a beam that is unsymmetrical and multi-layered, where the shear strain across the visco-elastic layer is not constant. Nakra (1978) provided a summary of damping treatments for structures including visco-elastic material use, multi-layered configuration and constrained layered configuration. Vinson and Sierakowski (2004) investigated the damping behavior of structures made of composite materials.

Initial research on the vibration of sandwich cylindrical shells was done by White (1960), Yu (1960), and Eienieck and Fraudenthal (1996), who included layers with visco-elastic properties. Jones and Salerno (1965) investigated the damping effect in the forced axisymmetric vibration of a cylindrical sandwich shell. The sandwich cylinders considered by Jones and Salerno (1977) were infinitely long or simply supported.

Pan (1968) studied the axisymmetrical vibrations of a circular sandwich shell using a visco-elastic layer with a complex shear modulus. He derived and presented the partial differential equations for the motion. The solution of these differential equations yields the frequency equation and the corresponding modal loss-factor after satisfying the boundary conditions. DiTaranto et al. (1973) provided experimental and theoretical data for a driving point impedance in damped composite rings. Lu et al. (1973) introduced damping material in the core layer and analyzed the mechanical impedance of damped three-layered sandwich rings by using the approximate method. They considered the effects of the operational temperature and frequency ranges on the visco-elastic material. The experimental data compared reasonably well with their theoretical predictions for lower modes. Lu and Everstine (1980) included a discontinuous layer in their presentation for the mechanical impedance of layered rings and beams. Hamidzadeh and Chandler (1991) analytically simulated one of the cases presented by Lu et al (1973) using forced vibration analysis. Their results compared satisfactorily well with those of Lu et al. (1973).

El-Rahed and Wanger (1986) considered the damping response of shells by using a constrained visco-elastic layer. Hamidzadeh et al. (1991) studied the damping of lobar vibrations in cylindrical structures. They developed an analytical solution for the constrained three-layer thick cylinders that has a visco-elastic middle layer. They utilized the stresses and displacements at all interfaces, and complied with the compatibility requirements at each interface. Hamidzadeh and Sawaya (1993) studied the effect of core material loss-factor, and the effect of the outer layer thickness on the natural frequencies and their corresponding modal loss-factor for a three-layer cylinder. Hamidzadeh and Jiang (1995) presented an analytical solution for the free vibration of thick elastic multi-layer cylinders. The presented method determined the natural frequencies and modal loss-factors for different circumferential wave numbers and a number of different geometric configurations. Hamidzadeh (2008) considered the effect of the visco-elastic core thickness on modal loss-factors of a thick three-layer cylinder.

1.9 Vibrations of Cylindrical Panels

An important step in the study of cylindrical panels is the evaluation of the modal displacement functions and their corresponding resonant frequencies. This modal information plays a key role in the design of cylindrical panels subjected to dynamic excitations. Most studies on the dynamic response of cylindrical panels have been restricted to the application of the theory of cylindrical shells. In the present theory, the equation of motion for a cylindrical shell is formed by taking into account the effect of transverse shear deformations and rotary inertia on the natural frequencies of the shell.

Hamidzadeh (1997) extended the research to develop the two-dimensional visco-elastic theory involving the cylindrical panels. The solution was developed to get the stresses and displacements for several panel angles, four-flexural mode, and a wide range of thickness modes. The resonant frequency results of cylindrical panels compared satisfactorily with those of complete cylinders. Hamidzadeh and Medhora (1995) considered the free transverse vibration of a composite thick shell with discontinues constraining layer.

2

Governing Equations

This chapter develops the governing equations of motion for a homogeneous isotropic elastic solid, using the linear three-dimensional theory of elasticity in cylindrical coordinates. At first, classical relationships between stress, strain, and displacement are reviewed and implemented into the dynamic equilibrium equations. The mathematical representations of the linear theory of elasticity derived in this chapter will set the stage for the development of the required governing equations for the possible modes of vibrations in cylindrical structures with any thickness.

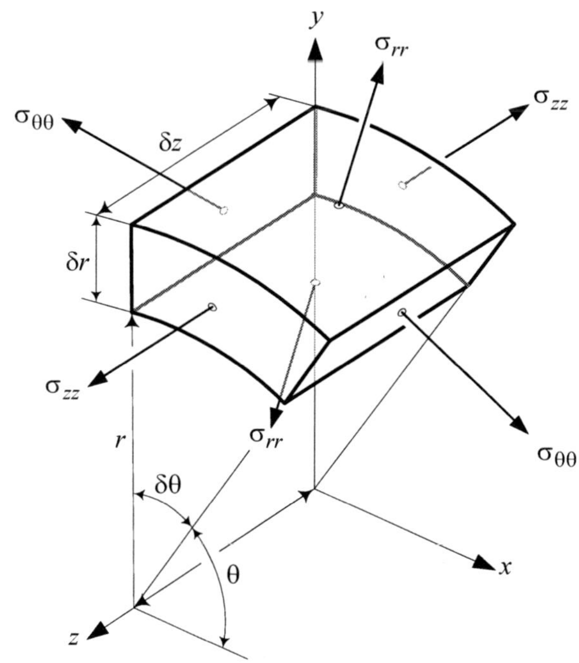

FIGURE 2.1. Direct Stresses in Cylindrical Coordinates.

A detailed mathematical review of elasto-dynamic problems can be found in most classical text books on advanced mechanics of materials and the theory of elasticity. In particular, references such as (Ford and Alexander 1963) and (Mal and Singh 1991) can provide the best extensive reviews

H.R. Hamidzadeh, R.N. Jazar, *Vibrations of Thick Cylindrical Structures*
DOI 10.1007/978-0-387-75591-5_2,

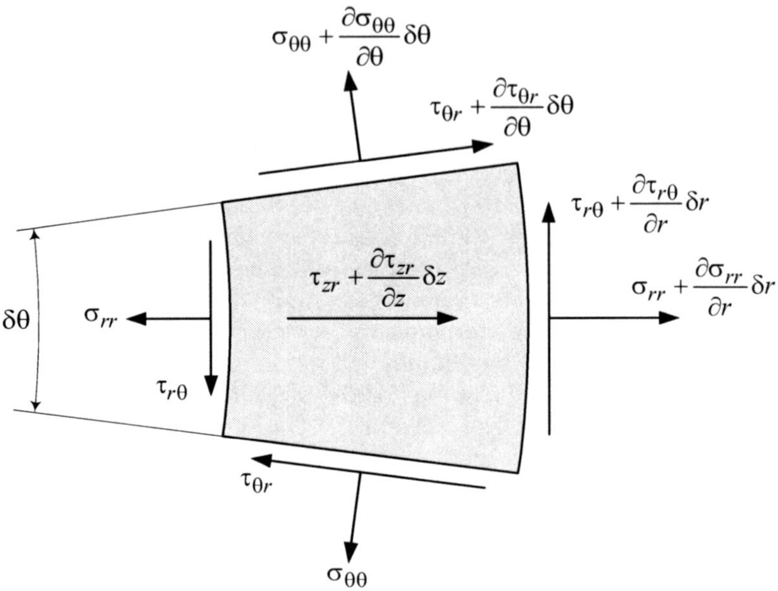

FIGURE 2.2. Stresses in the r and θ Directions.

of stress, strain, and displacement in cylindrical coordinates. The following sections provide a succinct review of essential topics needed for the establishment of the governing elasto-dynamic equations.

2.1 State of Stresses at a Point

A three dimensional state of stress in an infinitesimal cylindrical element is shown in the following three figures. Figure 2.1 depicts such an element with direct stresses, dimensions, and directions of the cylindrical coordinate. Figure 2.2 represents the direct and shear stresses in the radial and transverse directions (r and θ), and the variation of direct and shear stresses in these two directions. Figure 2.3 shows direct and shear stresses associated with the planes perpendicular to the r and z directions, as well as their variations along these directions.

In the above graphical representations the changes in direct and shear stresses are given by considering the first order infinitesimal term used in Taylor series approximation. The series approximation has been truncated after the second term. Further terms within the series representation contain terms of an infinitesimal length squared. Assuming that the second

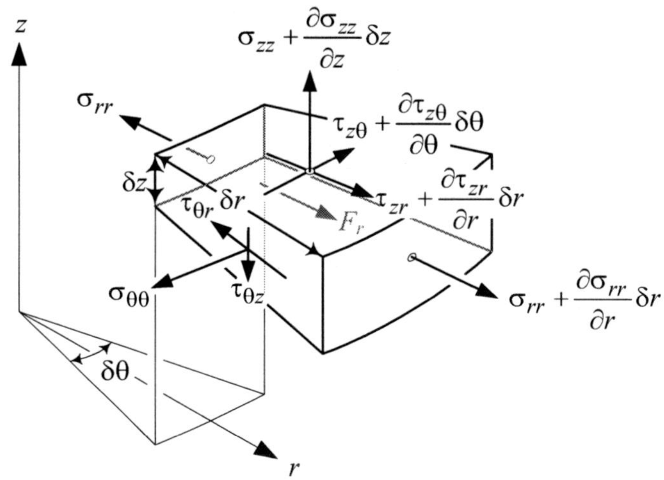

FIGURE 2.3. Stresses in the plane perpendicular to r and z direction.

order terms are very small, they can be neglected. Therefore, the change in stress across the element is considered very small.

2.2 Equilibrium Equations in Terms of Stress

Utilizing Newton's second law and the graphical representation of the state of stress, the equilibrium equations for an infinitesimal element in a cylindrical coordinates will be developed. By examining the state of stress on the element shown in section 2.1, the following equilibrium equation in the r direction is given.

$$\begin{aligned}
& \left(\sigma_{rr} + \frac{\partial \sigma_{rr}}{\partial r}\delta r\right)(r+\delta r)\,\delta\theta\delta z + \left(\tau_{r\theta} + \frac{\partial \tau_{r\theta}}{\partial \theta}\delta\theta\right)\delta r\delta z\cos\frac{\delta\theta}{2} \\
& + \left(\tau_{rz} + \frac{\partial \tau_{rz}}{\partial z}\delta z\right)\left(r + \frac{\delta r}{2}\right)\delta r\delta\theta + F_r\delta r\delta\theta\delta z \\
= \; & \sigma_{rr} r\delta\theta\delta z + \delta\tau_{r\theta}\delta r\delta z\cos\frac{\delta\theta}{2} + \tau_{rz}\left(r + \frac{\delta r}{2}\right)\delta r\delta\theta + \\
& + \left(\sigma_{\theta\theta} + \frac{\partial \sigma_{\theta\theta}}{\partial \theta}\delta\theta\right)\delta r\delta z\sin\frac{\delta\theta}{2} + \sigma_{\theta\theta}\delta r\delta z\sin\frac{\delta\theta}{2}
\end{aligned} \tag{2.1}$$

Canceling appropriate terms from both sides of the equation and after simplifying, it yields:

$$\frac{\partial \sigma_{rr}}{\partial r} + \frac{1}{r}\frac{\partial \tau_{r\theta}}{\partial \theta} + \frac{\partial \tau_{rz}}{\partial z} + \frac{\sigma_{rr} - \sigma_{\theta\theta}}{r} + F_r = 0 \tag{2.2}$$

Similarly, the equilibrium equation for the θ direction yields:

$$\frac{\partial \tau_{r\theta}}{\partial r} + \frac{1}{r}\frac{\partial \sigma_{\theta\theta}}{\partial \theta} + \frac{\partial \tau_{\theta z}}{\partial z} + \frac{2}{r}\tau_{r\theta} + F_\theta = 0 \tag{2.3}$$

and finally, for the z direction one may write:

$$\frac{\partial \tau_{rz}}{\partial r} + \frac{1}{r}\frac{\partial \tau_{\theta z}}{\partial \theta} + \frac{\partial \sigma_{zz}}{\partial z} + \frac{1}{r}\tau_{rz} + F_z = 0 \tag{2.4}$$

In the above simplifications, due to very small angle of $\delta\theta$, the following approximations were used:

$$\cos\frac{\delta\theta}{2} \approx 1 \qquad \sin\frac{\delta\theta}{2} \approx \frac{\delta\theta}{2} \tag{2.5}$$

In addition to the stresses, body forces acting throughout the element have been considered for each direction. These are denoted by F_r, F_θ, and F_z which are introduced as forces in the r, θ, and z direction per unit of volume. Due to the cancellation of the moments about each of the three perpendicular axes, the relations among the six shear stress components are presented by the following three equations:

$$\tau_{r\theta} = \tau_{\theta r} \qquad \tau_{\theta z} = \tau_{z\theta} \qquad \tau_{zr} = \tau_{rz} \tag{2.6}$$

Therefore, the stress at any point in the cylinder may be accurately described by three direct stresses and three shear stresses.

2.3 Stress-Strains Relationships

The constitutive relation between stresses and strains for a homogeneous and isotropic material can be expressed by Hooke's law. By definition, a homogeneous and isotropic material has the same properties in all directions. From this, the following three equations for direct strain in terms of stress are presented:

$$e_{rr}E = \sigma_{rr} - \nu\left(\sigma_{\theta\theta} + \sigma_{zz}\right) \tag{2.7}$$

$$e_{\theta\theta}E = \sigma_{\theta\theta} - \nu\left(\sigma_{zz} + \sigma_{\theta\theta}\right) \tag{2.8}$$

$$e_{zz}E = \sigma_{zz} - \nu\left(\sigma_{rr} + \sigma_{\theta\theta}\right) \tag{2.9}$$

where e_{rr}, $e_{\theta\theta}$, and e_{zz} are the direct strain in the r, θ, and z directions respectively; E is the *Young's modulus* or the modulus of elasticity; and ν is a proportionality factor called *Poisson's ratio.*

The other three Hooke's law relations result from the following proportionality between shear stresses and shear strains:

$$\tau_{r\theta} = Ge_{r\theta} \tag{2.10}$$

$$\tau_{rz} = Ge_{rz} \tag{2.11}$$

$$\tau_{\theta z} = Ge_{\theta z} \tag{2.12}$$

where, $e_{r\theta}$ is the shear strain along θ and perpendicular to r; e_{rz} is the shear strain along z and perpendicular to r; $e_{\theta z}$ is the shear strain along z and perpendicular to θ; and G is the *shear modulus* or the *modulus of rigidity.*

Through the general definition of shear stress and strain, the relationship between shear modulus, Young's modulus, and Poisson's ratio is given as:

$$G = \frac{E}{2(1+\nu)} \tag{2.13}$$

Lame's elastic constant, λ, and *volumetric strain,* ε are introduced by the following equations:

$$\lambda = \frac{\nu E}{(1-2\nu)(1+\nu)} \tag{2.14}$$

$$\varepsilon = e_{rr} + e_{\theta\theta} + e_{zz} \tag{2.15}$$

Then, a different form of Hooke's law relating direct stresses and direct strains can be achieved by adding the direct strain equations (2.7)-(2.9).

$$\frac{E\varepsilon}{1-2\nu} = \sigma_{rr} + \sigma_{\theta\theta} + \sigma_{zz} \tag{2.16}$$

Rearranging equation (2.7), one may write:

$$\sigma_{\theta\theta} + \sigma_{zz} = \frac{\sigma_{rr} - Ee_{rr}}{\nu} \tag{2.17}$$

Now, by substituting equation (2.17) into (2.16) it yields:

$$\frac{E\varepsilon}{1-2\nu} = \sigma_{rr} + \frac{\sigma_{rr}}{\nu} - \frac{Ee_{rr}}{\nu} \tag{2.18}$$

and the result can be arranged as:

$$\frac{\nu E\varepsilon}{1-2\nu} = (1+\nu)\,\sigma_{rr} - Ee_{rr} \tag{2.19}$$

from which the direct stress in the radial direction is determined to be:

$$\sigma_{rr} = \frac{\nu E}{(1 - 2\nu)(1 + \nu)} \varepsilon + \frac{E}{1 + \nu} e_{rr} \tag{2.20}$$

Now using the definitions of the shear modulus and Lame's elastic constant, the direct radial stress is presented as:

$$\sigma_{rr} = \lambda\varepsilon + 2Ge_{rr} \tag{2.21}$$

In a similar procedure, the direct circumferential stress and the direct axial stress are represented in terms of the volumetric strain, Lame's elastic constant, the shear modulus, and the appropriate direct strains are presented in the following equations.

$$\sigma_{\theta\theta} = \lambda\varepsilon + 2Ge_{\theta\theta} \tag{2.22}$$

$$\sigma_{zz} = \lambda\varepsilon + 2Ge_{zz} \tag{2.23}$$

2.4 Strain-Displacement Relationships

In Figure 2.4, a small element of an elastic homogenous and isotropic medium is represented in cylindrical coordinates. The element contains the point A, which represents a given point having the coordinates of (r, θ, z) and the point F, an infinitesimal distance away, having the coordinates $(r + \delta r,\ \theta + \delta\theta,\ z + \delta z)$. In this figure, the angle θ may be measured from any arbitrary coordinate direction such as x.

A typical small linear deformation of this element is depicted in Figure 2.5 where displacement and the distorted shape of the enlarged element is outlined. As can be seen, the displacement of point A to A' is defined by the three components of u_r, u_θ, and u_z. Where u_r, u_θ, and u_z are the displacements in the radial direction, transverse direction, and axial direction, respectively.

It should be noted that u_θ is the actual linear displacement along a circumferential arc. The displacements of the point F to $F\prime$ are $(u_r + \delta u_r)$, $(u_\theta + \delta u_\theta)$, and $(u_z + \delta u_z)$. Considering a horizontal plane, in Figure 2.6, the face $ACDB$ of the element previously shown in Figure 2.5 moves to $A'C'D'B'$ where there is a change in the length of the sides and the angles are sheared. Angle shearing is resulted by the change of the angle $\delta\theta$ to $(\delta\theta + \triangle\delta\theta)$.

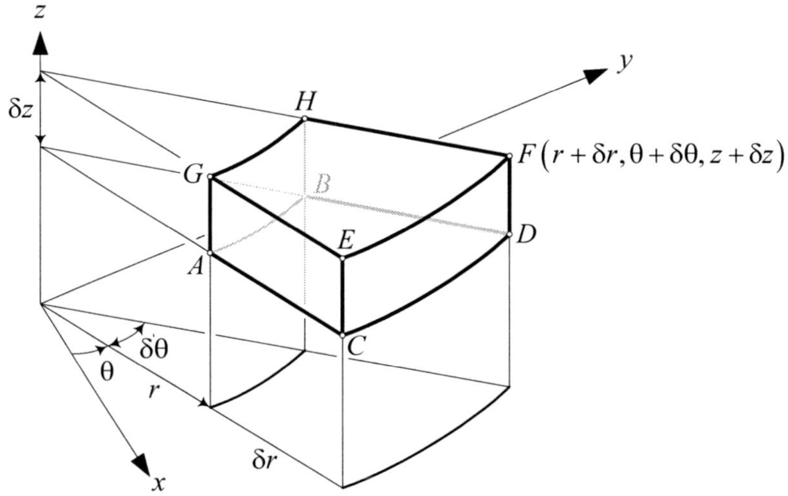

FIGURE 2.4. An Element in Cylindrical Coordinates.

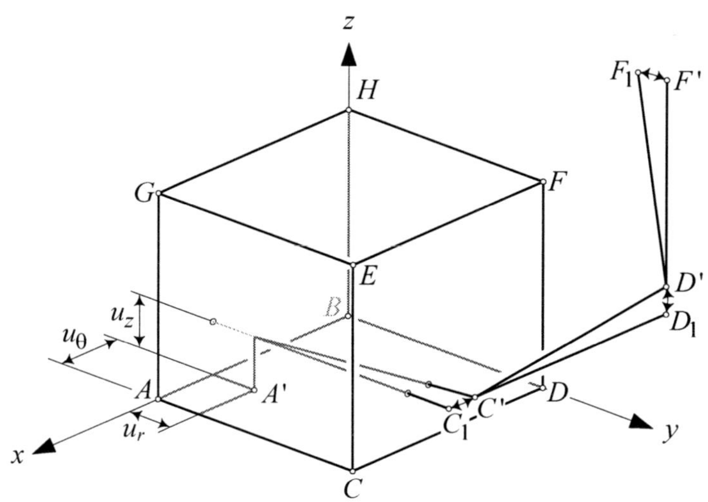

FIGURE 2.5. Element Subjected to Small Deformation.

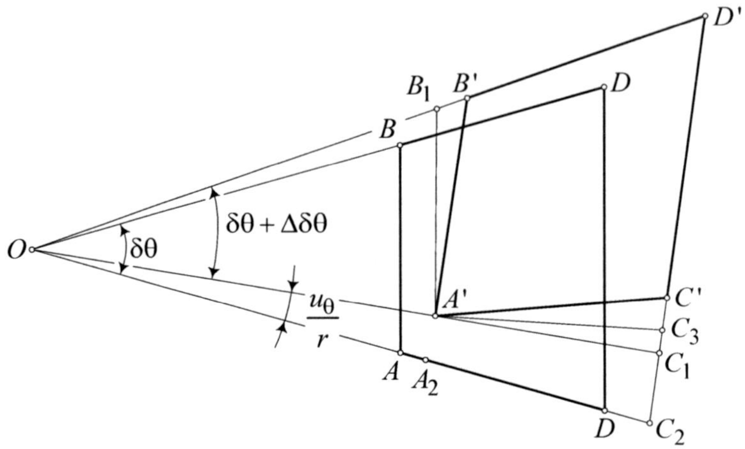

FIGURE 2.6. Horizontal Plane of Strained Element.

Upon examining the radial strain at point A and ignoring the effects of strains in the z direction, the strain in the side AC can be found. If the distance $A'C'$ is transferred to the line AC by drawing arcs, with center O, through A' and C' to intersect the line OAC at points A_2 and C_2 then the radial strain can be defined as:

$$e_{rr} = \frac{A_2C_2 - AC}{AC} \tag{2.24}$$

Considering the geometry of the Figure 2.6, the above equation can be written in the following form:

$$e_{rr} = \frac{\left(\delta r + \dfrac{\partial u_r}{\partial r}\delta r\right) - \delta r}{\delta r} = \frac{\partial u_r}{\partial r} \tag{2.25}$$

In a similar manner, the direct circumferential strain may be defined as:

$$e_{\theta\theta} = \frac{A'B' - AB}{AB} \tag{2.26}$$

where,

$$AB = r\delta\theta$$

and

$$A'B' = (r + u_r)\left(\delta\theta + \frac{\partial u_\theta}{r\partial\theta}\delta\theta\right). \tag{2.27}$$

In the definition of $A'B'$, the increase in the angle of the arc of $A'B'$ is given by $\delta u_\theta/(r + u_r)$ which can be approximated by $(\partial u_\theta/(r\partial\theta))\,\delta\theta$.

Therefore, by neglecting the second order terms, the circumferential direct strain is given by:

$$e_{\theta\theta} = \frac{\partial u_\theta}{r\partial\theta} + \frac{u_r}{r} \tag{2.28}$$

The shear strain, $e_{r\theta}$, is represented by the change of the angle BAC. By drawing $A'C_1$ parallel to AC, $A'B_1$ parallel to AB, and continuing line OA' to yield point C_3, the following procedures will yield $e_{r\theta}$ which is a rate of change of the line $A'C_1$. Notice that $A'C_1$ is parallel to line AC.

$$e_{r\theta} = \angle C_3A'C' + \angle B_1A'B' \tag{2.29}$$

where,

$$\angle C_3A'C' = \angle C_1A'C' - \angle C_1A'C_3 \tag{2.30}$$

and

$$\angle C_1A'C_3 = \angle AOA' = \frac{A_2A'}{r+u_r} = \frac{u_\theta}{r}. \tag{2.31}$$

The definition of the length C_1C' is:

$$C_1C' = \frac{\partial u_\theta}{\partial r}\delta r \tag{2.32}$$

then,

$$\angle C_1A'C' = \frac{C_1C'}{AC} = \frac{\partial u_\theta}{\partial r}. \tag{2.33}$$

Therefore, the shear angle $C_3A'C_1$ can be defined as:

$$\angle C_3A'C' = \frac{\frac{\partial u_\theta}{\partial r}\delta r}{\delta r} - \frac{u_\theta}{r} = \frac{\partial u_\theta}{\partial r} - \frac{u_\theta}{r} \tag{2.34}$$

The shear angle $B_1A'B'$ can be defined as:

$$\angle B_1A'B' = \frac{B_1B'}{r\delta\theta} \tag{2.35}$$

which is the radial displacement of B due to the angle $\delta\theta$ over the initial length. In partial derivative form this simplifies to:

$$\angle B_1A'B' = \frac{1}{r\delta\theta}\frac{\partial u_r}{\partial\theta}\delta\theta = \frac{\partial u_r}{r\partial\theta} \tag{2.36}$$

Therefore, the shear strain $e_{r\theta}$, is given as:

$$e_{r\theta} = \frac{\partial u_\theta}{\partial r} - \frac{u_\theta}{r} + \frac{\partial u_r}{r\partial\theta} \tag{2.37}$$

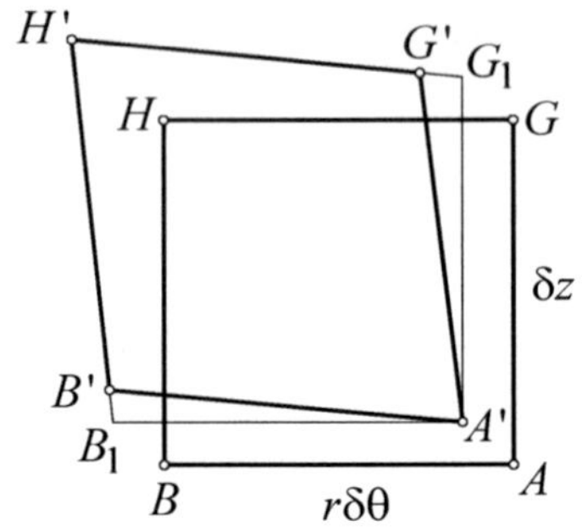

FIGURE 2.7. The (z, θ) Plane of Strained Element.

Considering the z direction, the direct axial strain can be defined similarly to the procedure used in Cartesian coordinates. Recall that the axial strain is defined as the ratio of the change in length to the original length of the element in the z direction. Examining the strain in line AF in the z direction, the direct axial strain is given as:

$$e_{zz} = \frac{A'_z F'_z - A_z F_z}{A_z F_z} \tag{2.38}$$

which simplifies to:

$$e_{zz} = \frac{\delta z + \frac{\partial u_z}{\partial z}\delta z - \delta z}{\delta z} \tag{2.39}$$

and finally to:

$$e_{zz} = \frac{\partial u_z}{\partial z} \tag{2.40}$$

In Figure 2.7 the z plane is shown as viewed from the origin. On the face $ABHG$ the shear strain $e_{\theta z}$ causes the right angle BAG to be displaced to $B'A'G'$. Note that $A'B_1$ is parallel to AB and $A'G'$ is parallel to AG. Therefore, the shear strain, $e_{\theta z}$, is given as:

$$e_{\theta z} = \angle G_1 A' G' + \angle B_1 A' B' \tag{2.41}$$

which is equivalent to:

$$e_{\theta z} = \frac{1}{\delta z}\frac{\partial u_\theta}{\partial z}\delta z + \frac{1}{r\delta\theta}\frac{\partial u_z}{\partial \theta}\delta\theta \tag{2.42}$$

and simplifies to:

$$e_{\theta z} = \frac{\partial u_\theta}{\partial z} + \frac{\partial u_z}{r\partial\theta} \tag{2.43}$$

Finally by examining the (r, z) plane, the shear strain, e_{rz}, is defined as:

$$e_{rz} = \angle G_1 A'G' + \angle C_1 A'C' \tag{2.44}$$

which yields:

$$e_{rz} = \frac{\frac{\partial u_r}{\partial z}\delta z}{\delta z} + \frac{\frac{\partial u_z}{\partial r}\delta r}{\delta r} \tag{2.45}$$

and simplifies to:

$$e_{rz} = \frac{\partial u_r}{\partial z} + \frac{\partial u_z}{\partial r} \tag{2.46}$$

There are now six strain components given in terms of the cylinder displacements. This completes the development of the required strain-displacement relationships.

2.5 Stress-Displacement Relationships

In this section, the stress-displacement relationships are developed by building upon Hooke's law and strain displacement relationships. Beginning with the direct radial stress in terms of strain and substituting the equations for direct strains, the radial stress in terms of displacement can be presented as:

$$\sigma_{rr} = \lambda\left(\frac{\partial u_r}{\partial r} + \frac{\partial u_\theta}{r\partial\theta} + \frac{u_r}{r} + \frac{\partial u_z}{\partial z}\right) + 2G\frac{\partial u_r}{\partial r} \tag{2.47}$$

The substituted direct strains are in terms of displacements and the volumetric strain. Similarly, the direct circumferential stress and direct axial stress, in terms of displacement, may be given as:

$$\sigma_{\theta\theta} = \lambda\left(\frac{\partial u_r}{\partial r} + \frac{\partial u_\theta}{r\partial\theta} + \frac{u_r}{r} + \frac{\partial u_z}{\partial z}\right) + 2G\left(\frac{\partial u_\theta}{r\partial\theta} + \frac{u_r}{r}\right) \tag{2.48}$$

$$\sigma_{zz} = \lambda\left(\frac{\partial u_r}{\partial r} + \frac{\partial u_\theta}{r\partial\theta} + \frac{u_r}{r} + \frac{\partial u_z}{\partial z}\right) + 2G\frac{\partial u_z}{\partial z} \tag{2.49}$$

Similarly, the three shear stresses in terms of shear strains are given by equations (2.10)-(2.12) and the shear strains, in terms of displacement components, are provided by equations (2.37), (2.43), and (2.46). Therefore, these shear stresses are:

$$\tau_{r\theta} = G\left(\frac{\partial u_\theta}{\partial r} - \frac{u_\theta}{r} + \frac{\partial u_r}{r\partial\theta}\right) \tag{2.50}$$

$$\tau_{\theta z} = G\left(\frac{\partial u_\theta}{\partial z} + \frac{\partial u_z}{r\partial\theta}\right) \tag{2.51}$$

$$\tau_{rz} = G\left(\frac{\partial u_r}{\partial z} + \frac{\partial u_z}{\partial r}\right) \tag{2.52}$$

2.6 Equations of Motion

In this section, the governing equations of motion in terms of a displacement vector are generated. The displacement vector is given as:

$$\mathbf{u} = u_r\hat{\imath}_r + u_\theta\hat{\imath}_\theta + u_z\hat{\imath}_z \tag{2.53}$$

where $\hat{\imath}_r$, $\hat{\imath}_\theta$, and $\hat{\imath}_z$ denote unit vectors directed along the (r, θ, and z) axes, respectively. Substituting Hooke's law equations into the dynamic equilibrium equations and introducing strain-displacement relationships yield the governing equations of motion:

$$\mu\boldsymbol{\nabla}^2 u_r + (\lambda+\mu)\frac{\partial\varepsilon}{\partial r} = \rho\frac{\partial^2 u_r}{\partial t^2} \tag{2.54}$$

$$\mu\boldsymbol{\nabla}^2 u_\theta + (\lambda+\mu)\frac{\partial\varepsilon}{r\partial\theta} = \rho\frac{\partial^2 u_\theta}{\partial t^2} \tag{2.55}$$

$$\mu\boldsymbol{\nabla}^2 u_z + (\lambda+\mu)\frac{\partial\varepsilon}{\partial z} = \rho\frac{\partial^2 u_z}{\partial t^2} \tag{2.56}$$

where μ is the same as shear modulus G, ε is the volumetric strain, and the $\boldsymbol{\nabla}^2$ is the three dimensional Laplacian operator in cylindrical coordinates defined by:

$$\boldsymbol{\nabla}^2 = \frac{\partial^2}{\partial r^2} + \frac{\partial}{r\partial r} + \frac{\partial^2}{r^2\partial\theta^2} + \frac{\partial^2}{\partial z^2} \tag{2.57}$$

Multiplying equation (2.54) by $\hat{\imath}_r$, equation (2.55) by $\hat{\imath}_\theta$, equation (2.56) by $\hat{\imath}_z$, and adding these three equations, the vector form of the governing equation of motion is given by:

$$\mu\boldsymbol{\nabla}^2\mathbf{u} + (\lambda+\mu)\,\boldsymbol{\nabla}\left(\boldsymbol{\nabla}\cdot\mathbf{u}\right) = \rho\frac{\partial^2\mathbf{u}}{\partial t^2} \tag{2.58}$$

where $\boldsymbol{\nabla}$ is the cylindrical gradient:

$$\boldsymbol{\nabla} = \left(\frac{\partial}{\partial r} + \frac{1}{r}\right)\hat{\imath}_r + \frac{\partial}{r\partial\theta}\hat{\imath}_\theta + \frac{\partial}{\partial z}\hat{\imath}_z \tag{2.59}$$

2.7 Key Symbols

$A, B, \cdots A', B'$	point label
e	strain
e_{rr}, $e_{\theta\theta}$, e_{zz}	direct strain in r, θ, z directions
$e_{r\theta}$	shear strain along θ and perpendicular to r
e_{rz}	shear strain along z and perpendicular to r
$e_{\theta z}$	shear strain along z and perpendicular to θ
E	Young's modulus
F_r, F_θ, F_z	forces in the r, θ, z directions per unit of volume
G, μ	shear modulus, modulus of rigidity
$\hat{\imath}_r$, $\hat{\imath}_\theta$, $\hat{\imath}_z$	unit vectors along the axes r, θ, z
r	radial direction
u_r, u_θ, u_z	displacements in r, θ, z directions
$\mathbf{u}$	displacement vector
z	axial direction
δ	variation
ε	volumetric strain
θ	transverse direction
λ	Lame's elastic constant
μ, G	shear modulus, modulus of rigidity
ν	Poisson's ratio
σ	normal stress
σ_{rr}, $\sigma_{\theta\theta}$, σ_{zz}	direct stress in the r, θ, z directions
τ	shear stress
$\tau_{r\theta}$	shear stress along θ and perpendicular to r
τ_{rz}	shear stress along z and perpendicular to r
$\tau_{\theta z}$	shear stress along z and perpendicular to θ
$\angle$	angle
$\boldsymbol{\nabla}^2$	Laplacian operator
$\boldsymbol{\nabla}$	gradient

3

Vibration of Single-Layer Cylinder

The following describes the procedures used to develop a mathematical model for the vibration analysis of a single-layer cylinder. The medium for the cylinder is assumed to be a homogeneous and isotropic elastic with infinite length. A finite length cylinder can only be directly analyzed for the cylinders with simply supported end conditions.

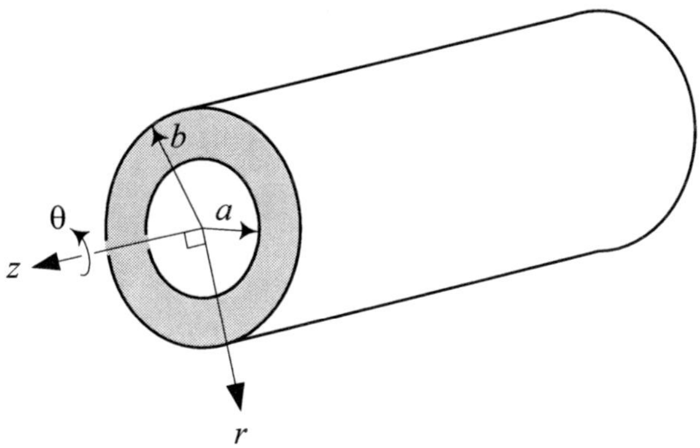

FIGURE 3.1. Reference Coordinates and Dimensions.

The presented mathematical procedure in this chapter is based on the work discussed by Armenakas et al. (1969), for the free vibration case. Their technique was refined by Hamidzadeh et al. (1981) to develop an analytical approach for studying the free, as well as, forced vibrations for these structures. The presented mathematical procedure implements the linear elasto-dynamic theory and formulates the displacements and stresses to satisfy the required boundary conditions for both cases of free and forced vibrations. Figure 3.1 describes the cylindrical coordinates and the required geometrical parameters.

H.R. Hamidzadeh, R.N. Jazar, *Vibrations of Thick Cylindrical Structures*
DOI 10.1007/978-0-387-75591-5_3,

3.1 Governing Equation

The governing equation of motion for a cylinder in invariant form was developed and presented in equation (2.58) in the previous chapter, and for continuity is repeated here.

$$\mu \boldsymbol{\nabla}^2 \mathbf{u} + (\lambda + \mu) \boldsymbol{\nabla} (\boldsymbol{\nabla} \cdot \mathbf{u}) = \rho \frac{\partial^2 \mathbf{u}}{\partial t^2} \tag{3.1}$$

where $\mathbf{u}$ is the displacements vector, ρ is the density of the medium, λ and μ are Lamé elastic constants, and $\boldsymbol{\nabla}^2$ is the three dimensional Laplacian operator.

The boundary conditions for these structures can be written in terms of boundary stresses in both inner and outer surfaces of the cylinder. These stress components are σ_{rr}, $\tau_{r\theta}$, τ_{rz}. They are applied at both the inner surface ($r = a$) and the outer surface ($r = b$) boundaries. Each of these boundary stresses is considered to be the source of excitations for forced vibration of the cylindrical structures.

3.2 Proposed Solution

The governing partial differential equations of motion can be solved using Lamé potential functions ϕ and $\mathbf{H}$, where the displacements vector $\mathbf{u}$ in terms of these potential functions is presented by the following equation.

$$\mathbf{u} = \boldsymbol{\nabla} \phi + \boldsymbol{\nabla} \times \mathbf{H} \tag{3.2}$$

where

$$\mathbf{H} = H_r \hat{\imath}_r + H_\theta \hat{\imath}_\theta + H_z \hat{\imath}_z . \tag{3.3}$$

In the following section, it will be shown that the above displacements vector given in terms of the Lamé potential functions ϕ and $\mathbf{H}$, will satisfy the governing equations of motion. Moreover, it will be proven that these potential functions can be presented using the following wave equations.

$$v_1^2 \, \boldsymbol{\nabla}^2 \phi = \frac{\partial^2 \phi}{\partial t^2} \tag{3.4}$$

$$v_1^2 \, \boldsymbol{\nabla}^2 \mathbf{H} = \frac{\partial^2 \mathbf{H}}{\partial t^2} \tag{3.5}$$

where, v_1 and v_2 are the velocities of propagation of the dilation and distortional wave, respectively. They can be presented as:

$$v_1 = \sqrt{\frac{\lambda + 2\mu}{\rho}} \tag{3.6}$$

$$v_2 = \sqrt{\frac{\mu}{\rho}} \tag{3.7}$$

3.3 Derivation of Wave Equation

By using Lamé potential functions, the governing equation of motion is reduced to two wave equations. These equations are developed in the following series of equations. First, substituting equation (3.2) into the governing equation of motion (3.1) yields:

$$\begin{aligned} &\mu \boldsymbol{\nabla}^2 (\boldsymbol{\nabla}\phi + \boldsymbol{\nabla} \times \mathbf{H}) + (\lambda + \mu) \boldsymbol{\nabla}\boldsymbol{\nabla} \cdot (\boldsymbol{\nabla}\phi + \boldsymbol{\nabla} \times \mathbf{H}) \\ = \ &\rho \frac{\partial^2}{\partial t^2} (\boldsymbol{\nabla}\phi + \boldsymbol{\nabla} \times \mathbf{H}) \end{aligned} \tag{3.8}$$

Expanding the above equation yields:

$$\begin{aligned} &\frac{\mu}{\rho}\boldsymbol{\nabla}^2\boldsymbol{\nabla}\phi + \frac{\mu}{\rho}\boldsymbol{\nabla}^2 (\boldsymbol{\nabla} \times \mathbf{H}) + \frac{\lambda}{\rho}\boldsymbol{\nabla}\boldsymbol{\nabla} \cdot \boldsymbol{\nabla}\phi + \frac{\mu}{\rho}\boldsymbol{\nabla}\boldsymbol{\nabla} \cdot \boldsymbol{\nabla}\phi \\ &\qquad + \frac{\lambda}{\rho}\boldsymbol{\nabla}\boldsymbol{\nabla} \cdot (\boldsymbol{\nabla} \times \mathbf{H}) + \frac{\mu}{\rho}\boldsymbol{\nabla}\boldsymbol{\nabla} \cdot (\boldsymbol{\nabla} \times \mathbf{H}) \\ &\qquad = \frac{\partial^2}{\partial t^2}\boldsymbol{\nabla}\phi + \frac{\partial^2}{\partial t^2} (\boldsymbol{\nabla} \times \mathbf{H}) \end{aligned} \tag{3.9}$$

Now, separating the terms containing the potential functions develops two independent equations in terms of ϕ and $\mathbf{H}$. Applying the property of $\boldsymbol{\nabla}^2 f = \boldsymbol{\nabla} \cdot \boldsymbol{\nabla} f$, the first equation can be simplified in the form:

$$\frac{\mu}{\rho}\boldsymbol{\nabla}\boldsymbol{\nabla}^2\phi + \frac{\lambda}{\rho}\boldsymbol{\nabla}\boldsymbol{\nabla}^2\phi + \frac{\mu}{\rho}\boldsymbol{\nabla}\boldsymbol{\nabla}^2\phi = \frac{\partial^2}{\partial t^2}\boldsymbol{\nabla}\phi \tag{3.10}$$

After rearranging and simplifying, the first wave equation will be written as:

$$\frac{\lambda + 2\mu}{\rho}\boldsymbol{\nabla}^2\phi = \frac{\partial^2 \phi}{\partial t^2} \tag{3.11}$$

Substituting equation (3.6) into the above equation yields:

$$v_1^2 \, \boldsymbol{\nabla}^2\phi = \frac{\partial^2 \phi}{\partial t^2} \tag{3.12}$$

The remainder of the equation (3.9) in terms of $\mathbf{H}$ can be simplified into the second wave equation.

$$\frac{\mu}{\rho}\boldsymbol{\nabla}^2(\boldsymbol{\nabla}\times\mathbf{H})+\frac{\lambda+\mu}{\rho}\boldsymbol{\nabla}\boldsymbol{\nabla}\cdot(\boldsymbol{\nabla}\times\mathbf{H})=\frac{\partial^2}{\partial t^2}(\boldsymbol{\nabla}\times\mathbf{H}) \tag{3.13}$$

Applying the vector property of $\boldsymbol{\nabla}\cdot(\boldsymbol{\nabla}\times\mathbf{V})=0$ (for any arbitrary vector of $\mathbf{V}$), the equation (3.13) reduces to:

$$\frac{\mu}{\rho}\boldsymbol{\nabla}^2(\boldsymbol{\nabla}\times\mathbf{H})=\frac{\partial^2}{\partial t^2}(\boldsymbol{\nabla}\times\mathbf{H}) \tag{3.14}$$

After simplifying and substituting equation (3.7) it yields:

$$v_2^2\boldsymbol{\nabla}^2\mathbf{H}=\frac{\partial^2}{\partial t^2}\mathbf{H} \tag{3.15}$$

This is the second wave equation. Considering Lamé potential functions, equations (3.12) and (3.15) are a modified form of the governing equations of motion.

3.4 Solutions to the Wave Equations

Throughout this section, the assumed solutions to the two modified governing equations (3.12) and (3.15) will be written in the following forms:

$$\phi = f(r)\cos(n\theta)\cos(\omega t+\zeta z) \tag{3.16}$$
$$H_r = g_r(r)\sin(n\theta)\sin(\omega t+\zeta z) \tag{3.17}$$
$$H_\theta = g_\theta(r)\cos(n\theta)\sin(\omega t+\zeta z) \tag{3.18}$$
$$H_z = g_3(r)\sin(n\theta)\cos(\omega t+\zeta z) \tag{3.19}$$

In the following sections it will be proven that the functions $f(r)$, $g_r(r)$, $g_\theta(r)$, and $g_3(r)$ are solutions to the following Bessel differential equations

$$B_{n,\alpha x}[f] = 0 \tag{3.20}$$
$$B_{n,\beta r}[g_3] = 0 \tag{3.21}$$
$$B_{n+1,\beta r}[g_r-g_\theta] = 0 \tag{3.22}$$
$$B_{n-1,\beta r}[g_r+g_\theta] = 0 \tag{3.23}$$

where the Bessel Differential Operator $B_{n,x}$ is defined as:

$$B_{n,x} = \frac{d^2}{dx^2} + \frac{d}{xdx} - \left(\frac{n^2}{x^2} - 1\right) \tag{3.24}$$

and n is an integer number.

3.4.1 Derivation of Bessel Equation $B_{n,\alpha x}[f] = 0$

The first wave equation was given in the following form:

$$v_1^2 \nabla^2 \phi - \frac{\partial^2}{\partial t^2}\phi = 0 \tag{3.25}$$

where:

$$\nabla^2 = \frac{\partial^2}{\partial r^2} + \frac{\partial}{r\partial r} + \frac{\partial^2}{r^2\partial\theta^2} + \frac{\partial^2}{\partial z^2} \tag{3.26}$$

$$\phi = f(r)\cos(n\theta)\cos(\omega t + \zeta z) \tag{3.27}$$

Substituting equations (3.26) and (3.27) into the first wave equation (3.25) and dividing by the harmonic term $\cos(n\theta)\cos(\omega t + \zeta z)$ yields:

$$v_1^2\left(\frac{d^2 f(r)}{dr^2} + \frac{df(r)}{rdr} - \left(\frac{n^2}{r^2} + \zeta^2\right) f(r)\right) + \omega^2 f(r) = 0 \tag{3.28}$$

Rearranging the terms yields:

$$\frac{d^2 f(r)}{dr^2} + \frac{df(r)}{rdr} - \frac{n^2 f(r)}{r^2} + \left(\frac{\omega^2}{v_1^2} - \zeta^2\right) f(r) = 0 \tag{3.29}$$

By defining α as:

$$\alpha^2 = \frac{\omega^2}{v_1^2} - \zeta^2 \tag{3.30}$$

and substituting into equation (3.29) yields:

$$\frac{d^2 f(r)}{dr^2} + \frac{df(r)}{rdr} - \frac{n^2 f(r)}{r^2} + \alpha^2 f(r) = 0 \tag{3.31}$$

Equation (3.31) can be arranged into the following two steps:

$$\frac{d^2 f(r)}{\alpha^2 dr^2} + \frac{df(r)}{\alpha^2 rdr} - \frac{n^2 f(r)}{\alpha^2 r^2} + f(r) = 0 \tag{3.32}$$

$$\frac{d^2 f(r)}{d(\alpha r)^2} + \frac{df(r)}{\alpha r d(\alpha r)} - \left(\frac{n^2}{(\alpha r)^2} - 1\right) f(r) = 0 \tag{3.33}$$

Here, equation (3.33) is the first Bessel differential equation, presented in equation (3.20).

3.4.2 Derivation of the Bessel Equation $B_{n,\beta r}\left[g_z\right]=0$

The second wave equation was presented by equation (3.15) and it can be separated into three scalar equations in terms of H_r, H_θ, and H_z. The scalar equation in terms of H_z can be presented as:

$$v_2^2\nabla^2 H_z-\frac{\partial^2}{\partial t^2}H_z=0 \tag{3.34}$$

where H_z is given by:

$$H_z=g_3\left(r\right)\sin\left(n\theta\right)\cos\left(\omega t+\zeta z\right) \tag{3.35}$$

Substituting H_z from (3.35) into (3.34) and dividing the results by the harmonic function $\sin\left(n\theta\right)\cos\left(\omega t+\zeta z\right)$ yields:

$$\frac{d^2g_3\left(r\right)}{dr^2}+\frac{dg_3\left(r\right)}{rdr}-\frac{n^2g_3\left(r\right)}{r^2}+\left(\frac{\omega^2}{v_2^2}-\zeta^2\right)g_3\left(r\right)=0 \tag{3.36}$$

By defining β as:

$$\beta^2=\frac{\omega^2}{v_2^2}-\zeta^2 \tag{3.37}$$

Substituting equation (3.37) into (3.36) after rearrangement and simplification yields:

$$\frac{d^2g_3\left(r\right)}{d\left(\beta r\right)^2}+\frac{dg_3\left(r\right)}{\beta rd\left(\beta r\right)}-\left(\frac{n^2}{\left(\beta r\right)^2}-1\right)g_3\left(r\right)=0 \tag{3.38}$$

Equation (3.38) is the second Bessel differential equation, which was presented by equation (3.21).

3.4.3 Derivation of the Bessel Equations $B_{n+1,\beta r}\left[g_r-g_\theta\right]=0$ and $B_{n-1,\beta r}\left[g_r+g_\theta\right]=0$

Let us begin with the second wave equation,

$$v_2^2\nabla^2\mathbf{H}-\frac{\partial^2}{\partial t^2}\mathbf{H}=0 \tag{3.39}$$

and reintroducing functions of H_r and H_θ as follows.

$$H_r = g_r\left(r\right)\sin\left(n\theta\right)\sin\left(\omega t+\zeta z\right) \tag{3.40}$$

$$H_\theta = g_\theta\left(r\right)\cos\left(n\theta\right)\sin\left(\omega t+\zeta z\right) \tag{3.41}$$

Then, substituting H_r into the wave equation (3.39) and dividing through by the harmonic term $\sin(n\theta)\sin(\omega t + \zeta z)$, after rearranging the result yields:

$$\frac{d^2 g_r(r)}{dr^2} + \frac{dg_r(r)}{rdr} - \frac{n^2 g_r(r)}{r^2} + \left(\frac{\omega^2}{v_2^2} - \zeta^2\right) g_r(r) = 0 \tag{3.42}$$

Introducing β from equation (3.37) to equation (3.42) yields a new Bessel differential equation.

$$\frac{d^2 g_r(r)}{d(\beta r)^2} + \frac{dg_r(r)}{\beta r d(\beta r)} - \left(\frac{n^2}{(\beta r)^2} - 1\right) g_r(r) = 0 \tag{3.43}$$

which can be presented in the following form:

$$B_{n,\beta r}\left[g_r(r)\right] = 0 \tag{3.44}$$

where, the constant n can be any integer value. Therefore, the following two equations are also acceptable:

$$B_{n+1,\beta r}\left[g_r(r)\right] = 0 \tag{3.45}$$
$$B_{n-1,\beta r}\left[g_r(r)\right] = 0 \tag{3.46}$$

Next, by substituting H_θ from equation (3.41) into the wave equation (3.39) and dividing through by the harmonic term yields:

$$v_2^2\left(\frac{d^2}{dr^2} + \frac{d}{rdr} - \frac{n^2}{r^2} - \zeta^2\right) g_\theta(r) + \omega^2 g_\theta(r) = 0 \tag{3.47}$$

Again, by introducing β from equation (3.37) to equation (3.47) yields an additional Bessel differential equation in the following form:

$$\frac{d^2 g_\theta(r)}{d(\beta r)^2} + \frac{dg_\theta(r)}{\beta r d(\beta r)} - \left(\frac{n^2}{(\beta r)^2} - 1\right) g_\theta(r) = 0 \tag{3.48}$$

This equation can be presented in the following Bessel Differential Operator form

$$B_{n,\beta r}\left[g_\theta(r)\right] = 0 \tag{3.49}$$

Again, n can be any integer value such as $n+1$ or $n-1$. Therefore, equation (3.49) can be written by the next two equations:

$$B_{n+1,\beta r}\left[g_\theta(r)\right] = 0 \tag{3.50}$$
$$B_{n-1,\beta r}\left[g_\theta(r)\right] = 0 \tag{3.51}$$

Subtracting equation (3.50) from equation (3.45) results in the third Bessel differential equation,

$$B_{n+1,\beta r}[g_r] - B_{n+1,\beta r}[g_\theta] = B_{n+1,\beta r}[g_r - g_\theta] = 0. \tag{3.52}$$

Adding equation (3.51) and (3.46) results in the fourth Bessel differential equation,

$$B_{n-1,\beta r}[g_r] + B_{n-1,\beta r}[g_\theta] = B_{n-1,\beta r}[g_r + g_\theta] = 0. \tag{3.53}$$

3.5 Solutions to the Bessel differential equations

The general solutions to the Bessel differential equations, (3.20)-(3.23), are discussed by many such as Watson (1962), Korenev (2002), Abramowitz and Stegun (1965), and Bell (1968). The solutions to the above mentioned differential equations are presented in the combination of the first and second kinds of Bessel functions by the following equations,

$$f(r) = A_1 J_n(\alpha r) + B_1 Y_n(\alpha r) \tag{3.54}$$

$$g_3(r) = A_2 J_n(\beta r) + B_2 Y_n(\beta r) \tag{3.55}$$

$$g_r(r) - g_\theta(r) = A_3 J_{n+1}(\beta r) + B_3 Y_{n+1}(\beta r) \tag{3.56}$$

$$g_r(r) + g_\theta(r) = A_4 J_{n-1}(\beta r) + B_4 Y_{n-1}(\beta r). \tag{3.57}$$

However, to simplify the equations for stresses and displacements, the following definitions are given:

$$2g_1(r) = g_r(r) - g_\theta(r) \tag{3.58}$$

$$2g_2(r) = g_r(r) + g_\theta(r) \tag{3.59}$$

It can be shown that any one of the three potential functions (i.e. g_1, g_2, or g_3) can be set equal to zero without the loss of the generality of the solutions, (Markus, 1988). Therefore, in this text the value of g_2 has been set to zero.

3.6 Displacements

Solving the resulted four wave equations, the solution to the Lamé potential functions ϕ, H_r, H_θ, and H_z are achieved. Furthermore, by substituting these potential functions into equation (3.2), the components of the displacement in the r, θ, z directions at any point within the medium of the cylindrical structure are determined by the following equations:

$$u_r = \left(f' + \frac{n}{r}g_3 + \zeta g_1\right)\cos(n\theta)\cos(\omega t + \zeta z) \tag{3.60}$$

$$u_\theta = \left(-\frac{n}{r}f + \zeta g_1 - g_3'\right)\sin(n\theta)\cos(\omega t + \zeta z) \tag{3.61}$$

$$u_z = \left(-\zeta f - g_1' - \frac{n+1}{r}g_1\right)\cos(n\theta)\sin(\omega t + \zeta z) \tag{3.62}$$

where, primes indicate differentiation with respect to r.

3.7 Stresses

The stress equations are developed by substituting the displacement equations into the stress-displacement relations. The resultant direct and shear stresses of interest are:

$$\sigma_{rr} = \left(-\lambda\left(\alpha^2 + \zeta^2\right)f + 2\mu\left(f'' + \frac{n}{r}\left(g_3' - \frac{g_3}{r}\right) + \zeta g_1'\right)\right)h_1 \tag{3.63}$$

$$\tau_{r\theta} = \left(-\frac{2n}{r}\left(f' - \frac{f}{r}\right) - \frac{n^2}{r^2}g_3 - g_3'' + \frac{g_3'}{r}\right.$$
$$\left. - \frac{n\zeta}{r}g_1 + \zeta g_1' - \frac{\zeta}{r}g_1\right)\mu h_2 \tag{3.64}$$

$$\tau_{rz} = \left(-2\zeta f' + \left(\frac{n+1}{r^2} - \zeta^2\right)g_1 - \frac{n+1}{r}g_1' - g_1'' - \frac{n\zeta}{r}g_3\right)\mu h_3 \tag{3.65}$$

where, primes indicate derivation with respect to r, and h_1, h_2, h_3 are defined by the following equations.

$$h_1 = \cos(n\theta)\cos(\omega t + \zeta z) \tag{3.66}$$

$$h_2 = \sin(n\theta)\cos(\omega t + \zeta z) \tag{3.67}$$

$$h_3 = \cos(n\theta)\sin(\omega t + \zeta z) \tag{3.68}$$

Although all six stress components can be derived in this form, nevertheless, only those stresses presented in equations (3.63)-(3.65) are needed for the analysis. In fact, for boundary stresses, only derivations of those stresses which occur on the inner and outer surfaces are presented in the next section.

3.7.1 Derivation of the Radial Stress σ_{rr}

This section outlines the development of the radial stress component which was introduced in equation (2.47) as:

$$\sigma_{rr} = \lambda \left(\frac{\partial u_r}{\partial r} + \frac{\partial u_\theta}{r\partial\theta} + \frac{u_r}{r} + \frac{\partial u_z}{\partial z} \right) + 2\mu \frac{\partial u_r}{\partial r} \tag{3.69}$$

Substituting the displacement equations (3.60)-(3.62), the radial stress can be presented as:

$$\begin{aligned} \frac{\sigma_{rr}}{h_1} &= \lambda \left(f'' + \frac{f'}{r} - \frac{n^2}{r^2} f - \zeta^2 f \right) \\ &\quad + 2\mu \left(f'' + \frac{n}{r} \left(g_3' - \frac{g_3}{r} \right) + \zeta g_1' \right) \end{aligned} \tag{3.70}$$

By adding and subtracting $\omega^2 f / v_1^2$ to the first bracket of equation (3.63), the above equation can be arranged into the following equation

$$\begin{aligned} \frac{\sigma_{rr}}{h_1} &= \lambda \left(f'' + \frac{f'}{r} - \frac{n^2}{r^2} f + \left(\frac{\omega^2}{v_1^2} - \zeta^2 \right) f - \frac{\omega^2}{v_1^2} f \right) \\ &\quad + 2\mu \left(f'' + \frac{n}{r} \left(g_3' - \frac{g_3}{r} \right) + \zeta g_1' \right) \end{aligned} \tag{3.71}$$

or

$$\begin{aligned} \frac{\sigma_{rr}}{h_1} &= \lambda \left(f'' + \frac{f'}{r} - \frac{n^2}{r^2} f + \alpha^2 f - \frac{\omega^2}{v_1^2} f \right) \\ &\quad + 2\mu \left(f'' + \frac{n}{r} \left(g_3' - \frac{g_3}{r} \right) + \zeta g_1' \right). \end{aligned} \tag{3.72}$$

Considering equation (3.24) and (3.20) the above equation will reduce to:

$$\frac{\sigma_{rr}}{h_1} = \lambda \left(B_{n,\alpha r} [f] - \frac{\omega^2}{v_1^2} f \right) + 2\mu \left(f'' + \frac{n}{r} \left(g_3' - \frac{g_3}{r} \right) + \zeta g_1' \right) \tag{3.73}$$

The Bessel differential equation in the first term is equal to zero. Therefore, equation (3.73) reduces to:

$$\frac{\sigma_{rr}}{h_1} = -\lambda \frac{\omega^2}{v_1^2} f + 2\mu \left(f'' + \frac{n}{r} \left(g_3' - \frac{g_3}{r} \right) + \zeta g_1' \right) \tag{3.74}$$

Substituting in terms of α and ζ for $(\omega/v_1)^2$ from (3.30) into the first term of equation (3.74), the final form of the direct radial stress is given as:

$$\begin{aligned} \sigma_{rr} &= -\lambda \left(\alpha^2 + \zeta^2 \right) f \cos(n\theta) \cos(\omega t + \zeta z) \\ &\quad + 2\mu \left(f'' + \frac{n}{r} \left(g_3' - \frac{g_3}{r} \right) + \zeta g_1' \right) \cos(n\theta) \cos(\omega t + \zeta z) \end{aligned} \tag{3.75}$$

where, prime indicates differentiation with respect to r.

3.7.2 Derivation of the Shear Stress $\tau_{r\theta}$

The tangential shear stress in terms of the displacements was presented by equation (2.50) as:

$$\tau_{r\theta} = \mu\left(\frac{\partial u_r}{r\partial\theta} + \frac{\partial u_\theta}{\partial r} - \frac{u_\theta}{r}\right) \tag{3.76}$$

Substitution of the displacement equations (3.60)-(3.62) into the above equation yields:

$$\begin{aligned}\frac{\tau_{r\theta}}{h_2} &= -\mu\frac{n}{r}\left(f' + \frac{n}{r}g_3 + \zeta g_1\right) + \mu\left(\frac{n}{r^2}f - \frac{n}{r}f' + \zeta g_1' + g_3''\right)\\ &\quad -\mu\left(-\frac{n}{r^2}f + \frac{\zeta}{r}g_1 - \frac{g_3'}{r}\right)\end{aligned} \tag{3.77}$$

Common terms are grouped and the final form of the tangential shear stress $\tau_{r\theta}$ is presented by:

$$\begin{aligned}\frac{\tau_{r\theta}}{h_2} &= -\mu\frac{2n}{r}\left(f' - \frac{f}{r}\right) + \mu\left(-\frac{n^2}{r^2}g_3 - g_3'' + \frac{g_3'}{r}\right)\\ &\quad +\mu\left(-\frac{n\zeta}{r}g_1 + \zeta g_1' - \frac{\zeta}{r}g_1\right)\end{aligned} \tag{3.78}$$

where, prime indicates differentiation with respect to r.

3.7.3 Derivation of the Shear Stress τ_{rz}

The tangential shear stress τ_{rz} was presented by the equation (2.52) as:

$$\tau_{rz} = \mu\left(\frac{\partial u_r}{\partial z} + \frac{\partial u_z}{\partial r}\right) \tag{3.79}$$

Substituting the displacement equations (3.60)-(3.62), the expanded form of equation (3.79) yields:

$$\begin{aligned}\frac{\tau_{rz}}{h_3} &= -\mu\zeta\left(f' + \frac{n}{r}g_3 + g_1\right)\\ &\quad +\mu\left(-\zeta f' - g_1'' + \frac{n+1}{r^2}g_1 - \frac{n+1}{r^2}g_1'\right)\end{aligned} \tag{3.80}$$

Grouping common terms, simplifies the tangential shear stress τ_{rz} to:

$$\begin{aligned}\frac{\tau_{rz}}{h_3} &= \mu\left(-2\zeta f' + \left(\frac{n+1}{r^2} - \zeta^2\right)g_1\right)\\ &\quad +\mu\left(-\frac{n+1}{r^2}g_1' - g_1'' - \frac{n\zeta}{r}g_3\right)\end{aligned} \tag{3.81}$$

3.8 The Boundary Stresses

By entering the solutions to the Bessel differential equations into the three stress components, equations (3.63)-(3.65) can be written in terms of the Bessel functions and their constant coefficients. Assuming arbitrary boundary condition values at the inner and outer surfaces of the cylinder, a linear non-homogeneous system of equations is developed. The matrix form of this system can be expressed by the following relation:

$$T\mathbf{X} = \mathbf{S} \tag{3.82}$$

where, $\mathbf{X}$ is:

$$\mathbf{X} = \begin{Bmatrix} A_1 \\ B_1 \\ A_2 \\ B_2 \\ A_3 \\ B_3 \end{Bmatrix} \tag{3.83}$$

and $\mathbf{S}$ is the vector of boundary stresses presented by the following equation:

$$\mathbf{S} = \begin{Bmatrix} \sigma_{rr}\left(r = a\right)/h_1 \\ \tau_{r\theta}\left(r = a\right)/h_2 \\ \tau_{rz}\left(r = a\right)/h_3 \\ \sigma_{rr}\left(r = b\right)/h_1 \\ \tau_{r\theta}\left(r = b\right)/h_2 \\ \tau_{rz}\left(r = b\right)/h_3 \end{Bmatrix} \tag{3.84}$$

The elements in the first three rows of the T matrix are developed by substituting $r = a$ for the inner boundary radius in equations (3.63)-(3.65). These elements are:

$$T_{1,1} = -\lambda\left(\alpha^2 + \zeta^2\right) Z_n\left(\alpha r\right) + 2\mu\alpha^2 Z_n''\left(\alpha r\right) \tag{3.85}$$

$$T_{1,2} = -\lambda\left(\alpha^2 + \zeta^2\right) W_n\left(\alpha r\right) + 2\mu\alpha^2 W_n''\left(\alpha r\right) \tag{3.86}$$

$$T_{1,3} = 2\mu\zeta\beta Z_{n+1}'\left(\beta r\right) \tag{3.87}$$

$$T_{1,4} = 2\mu\zeta\beta W_{n+1}'\left(\beta r\right) \tag{3.88}$$

$$T_{1,5} = 2\mu\frac{n}{r}\beta Z_n'\left(\beta r\right) - 2\mu\frac{n}{r^2} Z_n\left(\beta r\right) \tag{3.89}$$

$$T_{1,6} = 2\mu\frac{n}{r}\beta W_n'\left(\beta r\right) - 2\mu\frac{n}{r^2} W_n\left(\beta r\right) \tag{3.90}$$

$$T_{2,1} = -2\mu\frac{n}{r}\alpha Z_n'\left(\alpha r\right) + 2\mu\frac{n}{r^2} Z_n\left(\alpha r\right) \tag{3.91}$$

$$T_{2,2} = -2\mu\frac{n}{r}\alpha W_n'\left(\alpha r\right) + 2\mu\frac{n}{r^2} W_n\left(\alpha r\right) \tag{3.92}$$

$$T_{2,3} = -\mu\zeta\frac{n+1}{r} Z_{n+1}\left(\beta r\right) + \mu\zeta\beta W_{n+1}'\left(\beta r\right) \tag{3.93}$$

$$T_{2,4} = -\mu\zeta\frac{n+1}{r}W_{n+1}(\beta r) + \mu\zeta\beta W'_{n+1}(\beta r) \quad (3.94)$$

$$T_{2,5} = -\mu\frac{n^2}{r^2}Z_n(\beta r) - \mu\beta^2 Z''_n(\beta r) + \frac{\mu}{r}\beta Z'_n(\beta r) \quad (3.95)$$

$$T_{2,6} = -\mu\frac{n^2}{r^2}W_n(\beta r) - \mu\beta^2 W''_n(\beta r) + \frac{\mu}{r}\beta W'_n(\beta r) \quad (3.96)$$

$$T_{3,1} = -2\mu\zeta\alpha Z'_n(\alpha r) \quad (3.97)$$

$$T_{3,2} = -2\mu\zeta\alpha W'_n(\alpha r) \quad (3.98)$$

$$T_{3,3} = \mu\left(\frac{n+1}{r^2} - \zeta^2\right)Z_{n+1}(\beta r) - \mu\frac{n+1}{r}\beta Z'_{n+1}(\beta r) - \mu\beta^2 Z''_{n+1}(\beta r) \quad (3.99)$$

$$T_{3,4} = \mu\left(\frac{n+1}{r^2} - \zeta^2\right)W_{n+1}(\beta r) - \mu\frac{n+1}{r}\beta W'_{n+1}(\beta r) - \mu\beta^2 W''_{n+1}(\beta r) \quad (3.100)$$

$$T_{3,5} = -\mu\zeta\frac{n}{r}Z_n(\beta r) \quad (3.101)$$

$$T_{3,6} = -\mu\zeta\frac{n}{r}W_n(\beta r) \quad (3.102)$$

where primes denote differentiation with respect to the argument of the Bessel function. Also, Z_n and W_n denote the Bessel functions J_n and Y_n or the modified Bessel functions I_n and K_n. The type of Bessel function Z_n and W_n represent, depends on whether the argument is real or imaginary. The difference in sign between the recursion formula of the Bessel functions and the modified Bessel functions is handled within the Fortran code. Therefore, no modification of the given T matrix is needed. The proper selection of Bessel functions is given in table 3.1.

Table 3.1 - Bessel functions for Different Intervals of the Frequency.

Interval	Functions Used
$\omega > v_1\zeta$	$J(\alpha r), Y(\alpha r), J(\beta r), Y(\beta r)$
$v_1\zeta > \omega > v_2\zeta$	$I(\alpha r), K(\alpha r), J(\beta r), Y(\beta r)$
$\omega < v_2\zeta$	$I(\alpha r), K(\alpha r), I(\beta r), K(\beta r)$

Conversion of Bessel function derivatives is conducted according to the following steps.

$$\frac{d}{dr}f = \frac{d}{dr}\left(A_1 J_n(\alpha r) + B_1 Y_n(\alpha r)\right) \quad (3.103)$$

$$\frac{d}{dr}f = \alpha A_1\frac{d}{\alpha dr}J_n(\alpha r) + \alpha B_1\frac{d}{\alpha dr}Y_n(\alpha r) \quad (3.104)$$

$$\frac{d}{dr}f = \alpha A_1\frac{d}{d(\alpha r)}J_n(\alpha r) + \alpha B_1\frac{d}{d(\alpha r)}Y_n(\alpha r) \quad (3.105)$$

The last three rows of the T matrix are formed by substituting $r = b$ for the outer boundary radius. In the case of free vibration, the boundary stresses are zero. Therefore, the determinant of the matrix T must be zero. By solving the resulted equation, the natural frequencies for all modes of free vibration will be determined.

In the case of forced vibration, when the known modal boundary stress components are input into equation (3.82), the vector $\mathbf{X}$ is calculated. The vector $\mathbf{X}$ is then substituted into equations (3.60)-(3.65) to determine all six modal displacements and stresses at any point within the medium.

3.9 The Stress-Displacement Matrix Equation

From the stress matrix and the known boundary conditions, the coefficient vector $\mathbf{X}$ can be determined, as shown by the following equation:

$$\mathbf{X} = T^{-1}\mathbf{S} \tag{3.106}$$

Utilizing the known displacement and stress functions from equations (3.60)-(3.65) and substituting the Bessel differential equation solutions, a new system of linear equations can be built. The matrix form of this system can be presented by:

$$D\mathbf{X} = \mathbf{C} \tag{3.107}$$

where the vector $\mathbf{X}$ is known from the equation (3.106). The unknown vector $\mathbf{C}$ is defined as:

$$\mathbf{C} = \begin{Bmatrix} u_r\left(r=a\right)/h_1 \\ u_\theta\left(r=a\right)/h_2 \\ u_z\left(r=a\right)/h_3 \\ u_r\left(r=b\right)/h_1 \\ u_\theta\left(r=b\right)/h_2 \\ u_z\left(r=b\right)/h_3 \end{Bmatrix} \tag{3.108}$$

The vector $\mathbf{C}$ contains the resultant modal stresses and displacements of the cylinder. The first three rows of the D matrix, which are composed of the displacement equations, are given as follows:

$$D_{1,1} = \alpha Z_n'\left(\alpha r\right) \tag{3.109}$$

$$D_{1,2} = \alpha W_n'\left(\alpha r\right) \tag{3.110}$$

$$D_{1,3} = \alpha Z_{n+1}\left(\beta r\right) \tag{3.111}$$

$$D_{1,4} = \alpha W_{n+1}(\beta r) \tag{3.112}$$
$$D_{1,5} = \frac{n}{r} Z_n(\beta r) \tag{3.113}$$
$$D_{1,6} = \frac{n}{r} W_n(\beta r) \tag{3.114}$$

$$D_{2,1} = -\frac{n}{r} Z_n(\alpha r) \tag{3.115}$$
$$D_{2,2} = -\frac{n}{r} W_n(\alpha r) \tag{3.116}$$
$$D_{2,3} = \zeta Z_{n+1}(\beta r) \tag{3.117}$$

$$D_{2,4} = \zeta W_{n+1}(\beta r) \tag{3.118}$$
$$D_{2,5} = -\beta Z'_n(\beta r) \tag{3.119}$$
$$D_{2,6} = -\beta W'_n(\beta r) \tag{3.120}$$

$$D_{3,1} = -\zeta Z_n(\alpha r) \tag{3.121}$$
$$D_{3,2} = -\zeta W_n(\alpha r) \tag{3.122}$$
$$D_{3,3} = -\beta Z'_{n+1}(\beta r) - \frac{n+1}{r} Z_{n+1}(\beta r) \tag{3.123}$$

$$D_{3,4} = -\beta W'_{n+1}(\beta r) - \frac{n+1}{r} W_{n+1}(\beta r) \tag{3.124}$$
$$D_{3,5} = 0 \tag{3.125}$$
$$D_{3,6} = 0 \tag{3.126}$$

Again, the appropriate functions to be used in place of W and Z are given in table 3.1. The last three rows of the D matrix are identical to the last three rows of the T matrix. The solution to equation (3.107), vector $\mathbf{C}$, can yield displacements and stresses at any point within the cylinder.

3.10 Governing Equations for Free and Forced Vibrations

Free vibration analysis requires free boundary stresses at the inner and outer radii of $r = a$ and $r = b$. Thus, elements of vector $\mathbf{S}$ in equation (3.106) are zero. The nontrivial solution to the values of vector $\mathbf{X}$ in that equation requires that the determinant of the matrix T become zero.

$$|T| = 0 \tag{3.127}$$

This is the frequency equation which has an infinite number of solutions for the frequency of ω for every given integer number for circumferential wave number n, and real value for the axial wave number ζ. After determining the nontrivial elements of vector $\mathbf{X}$ for any specific values of n and ζ and by substituting them into equation (3.107), modal displacements and stresses for any given radius can be determined.

The governing equation for the forced vibration of thick cylindrical structure, excited by given harmonic boundary stresses, can be obtained by determining the vector $\mathbf{X}$ from equation (3.82) in terms of boundary stresses as presented in the following equation:

$$\mathbf{X} = T^{-1}\mathbf{S} \tag{3.128}$$

The harmonic response of displacements and stresses on any point on a surface perpendicular to radial direction can be presented by substituting vector $\mathbf{X}$ from equation (3.128) into equation (3.107).

$$\mathbf{C} = DT^{-1}\mathbf{S} \tag{3.129}$$

The frequency response for these displacements and stresses can be achieved by computing vector $\mathbf{C}$ for a given range of exciting frequencies.

3.11 Key Symbols

$B_{n,x}$	Bessel Differential Operator
$\mathbf{C}$	vector of boundary displacements, 6×1
D	matrix of Bessel functions, 6×6
e	strain
E	Young's modulus
$f(r)$	solution of Bessel equation $B_{n,\alpha x}[f] = 0$
g_1, g_2, g_3	potential functions
$g_r(r)$	solution of Bessel equation $B_{n,\beta r}[g_r] = 0$
$g_\theta(r)$	solution of Bessel equation $B_{n,\beta r}[g_\theta] = 0$
$g_3(r)$	solution of Bessel equation $B_{n,\beta r}[g_3] = 0$
G, μ	shear modulus, modulus of rigidity
$\mathbf{H}$,ϕ	Lamé potential functions
$\hat{\imath}_r$, $\hat{\imath}_\theta$, $\hat{\imath}_z$	unit vectors along the axes r, θ, z
J_n, Y_n	Bessel functions
n	integer number
r	radial direction
$r = a$	inner surface boundary
$r = b$	outer surface boundary
S	vector of boundary stresses, 6×1
t	time
T	matrix of Bessel functions, 6×6
u_r, u_θ, u_z	displacements in r, θ, z directions
$\mathbf{u}$	displacement vector
v_1	velocity of propagation of the dilation wave
v_2	velocity of propagation of the distortional wave
$\mathbf{X}$	vector of constants, vector of unknowns, 6×1
z	axial direction
W_n, Z_n	Bessel functions
I_n, K_n	modified Bessel functions
α^2	$= \omega^2/v_1^2 - \zeta^2$
β^2	$= \omega^2/v_2^2 - \zeta^2$
δ	variation
ε	volumetric strain
θ	transverse direction
λ	Lame's elastic constant
μ, G	shear modulus, modulus of rigidity
ν	Poisson's ratio
ρ	density
σ	normal stress
σ_{rr}, $\sigma_{\theta\theta}$, σ_{zz}	direct stress in the r, θ, z directions

τ	shear stress
$\tau_{r\theta}$	shear stress along θ and perpendicular to r
τ_{rz}	shear stress along z and perpendicular to r
$\tau_{\theta z}$	shear stress along z and perpendicular to θ
ϕ, $\mathbf{H}$	Lamé potential functions
ω	frequency
ζ	axial wave number
$'$	differentiation of Bessel function with respect to its argument
$\boldsymbol{\nabla}^2$	Laplacian operator
$\boldsymbol{\nabla}$	gradient

4

Modal Analysis of Cylindrical Structures

An overview, providing a physical understanding of the vibration modes for a single layer thick cylindrical structure, the main variables, and their general effect upon the response of the system, is given in this chapter. Furthermore, non-dimensional natural frequencies for a number of ratios of thickness to mean radius, different non-dimensional axial wavelengths, and circumferential wave numbers are tabulated and their respective modes are identified.

4.1 Frequency Factor

To simplify the presentation of results and comparison with other established results for specific cases, natural and resonant frequencies are normalized and introduced as the frequency factor, which is defined by the following equation:

$$\Omega = \frac{\omega}{\omega_s} \tag{4.1}$$

where,

$$\omega_s = \frac{\pi v_2}{H}. \tag{4.2}$$

ω_s is defined as the lowest simple thickness shear frequency of an infinite plate of thickness H with the same elastic constants as those of the cylinder. The parameter ω is the excitation frequency, in units of rad/s. For the purpose of comparing of results, the frequency factor is the same as the one defined by Armenakas et al. (1969).

4.2 Geometry of the Cylinder

The definitions of the various geometric parameters are evident in figure 3.1. However, a more thorough description is given here. The variables a and b indicate the inner and outer radii of the cylinder, respectively. H is

H.R. Hamidzadeh, R.N. Jazar, *Vibrations of Thick Cylindrical Structures*
DOI 10.1007/978-0-387-75591-5_4, © Springer Science+Business Media, LLC 2010

defined as the thickness of the cylinder and R is the mean radius of the cylinder.

In the numerical computation, a range of $H/R = 0.0$ to 2.0 was considered. The case of $H/R = 2.0$ approximates a solid cylinder. L is defined as the axial half wavelength.

4.3 Properties of the Medium

The medium and dynamic behavior of single layer cylindrical structures depend on their elastic characteristics. These characteristics are redefined here to summarize needed equations. ρ is the density of the material. ν is defined as Poisson's ratio. μ and λ are Lamé elastic constants which are defined as:

$$\mu = G \tag{4.3}$$

and

$$\lambda = \frac{2\nu}{1-2\nu} G. \tag{4.4}$$

For the homogeneous and isotropic materials, the shear modulus is related to the elastic modulus and the Poisson's ratio as shown by the following relationship.

$$G = \frac{E}{2\left(1-\nu\right)} \tag{4.5}$$

Assuming that the material for the cylinder is a linear visco-elastic with a constant material loss-factor, both the elastic and shear moduli are presented as complex numbers,

$$E = E'\left(1+i\eta\right) \tag{4.6}$$

$$G = G'\left(1+i\eta\right) \tag{4.7}$$

where, G' and E' are real parts of G and E respectively. η is defined as the loss-factor of the material.

4.4 Axial Wavelength

The axial wavelength is dictated by the wavelength of the applied modal boundary stresses. Considering simply supported end conditions for the

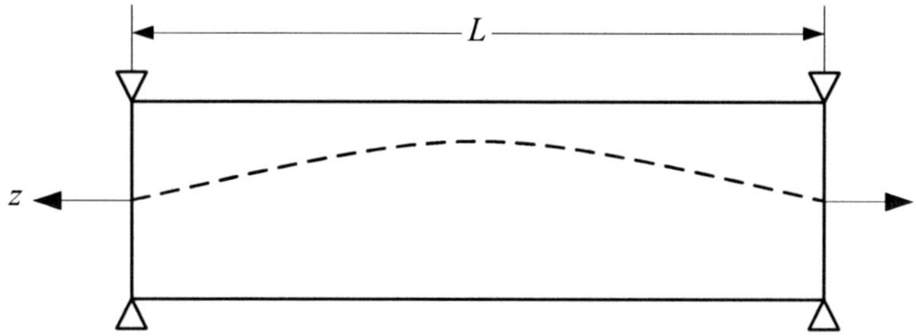

FIGURE 4.1. Fundamental Axial Mode.

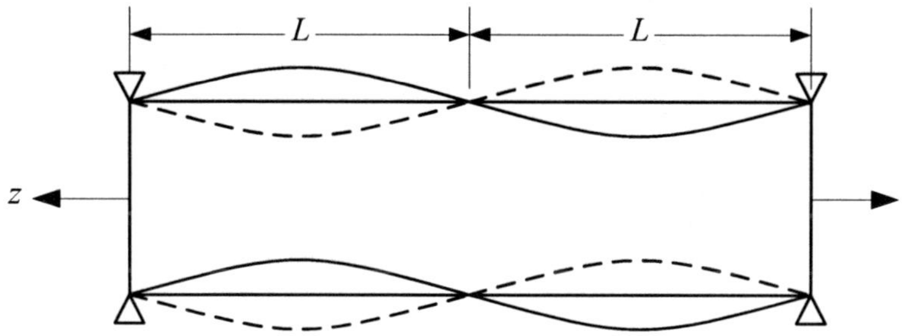

FIGURE 4.2. Second Axial Mode.

finite length cylinder case, the axial wave number, ζ, can be found with the following equation:

$$\zeta = \frac{\pi}{L} \tag{4.8}$$

Recall that the variable L is defined as 1/2 of the axial wavelength. Figures 4.1 and 4.2 show the axial half wavelength for the fundamental and second longitudinal modes, respectively. For computation, the values of H/L were input from approximately 0.0 to 1.0. $H/L = 0.0$ corresponds to the infinite wavelength case. $H/L > 0.0$ corresponds to finite length cylinders with simply supported end conditions. Having the thickness of the cylinder and the length of L, the respective values of H/L and ζ are calculated for determination of the natural frequencies and resonant frequencies.

4.5 Modes of Vibration

Depending on the value of nodal diameter (n), the axial wavelength (ξ), and specific values of $f(r)$, $g_r(r)$, $g_\theta(r)$, and $g_3(r)$ in equations (3.16)-(3.19)

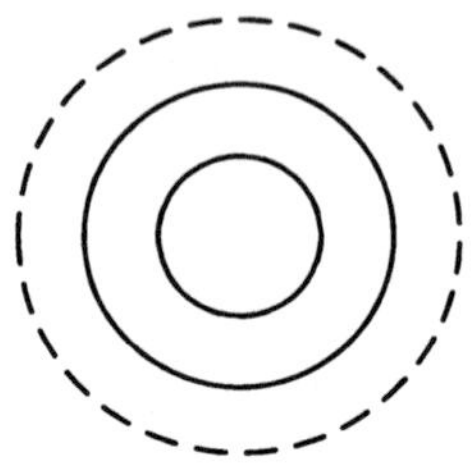

FIGURE 4.3. Breathing Mode $n = 0$.

infinite numbers of vibration modes can occur. In the following subsections, these modes are described.

4.5.1 Modes with Different Circumferential Wave Number

These modes can be categorized in the following groups.

1. breathing mode for $n = 0$ (where any point on the cross section of the cylinder vibrates harmonically in its radial direction)
2. torsional mode for $n = 0$ (associated with transverse displacement of a long circumference of the cylinder)
3. axial mode for $n = 0$ (associated with axial vibration)
4. bending and axial shear modes for $n = 1$ (where the cross section of the cylinder remains the same)
5. lobar modes for $n > 1$ (where both radial and transverse displacements vary with $\sin(n\theta)$)

By superposition, the displacement response of a single layer cylinder to an arbitrarily excited cylinder is composed of combinations of the infinite number of modes. Nevertheless, each mode can be excited by their respective modal boundary stresses.

In this chapter, a number of individual mode shapes for cross sections of the cylinders are illustrated in Figures 4.3-4.8. The lower modes ($n = 0, 1, 2, 3$) have been found to be the dominant modes of vibration. In fact, analysis of the motion for a point in a cylinder can be approximated by summing the effects of the dominant modes. Therefore, addition of higher lobar modes yields a more accurate analysis of the cylinder's response.

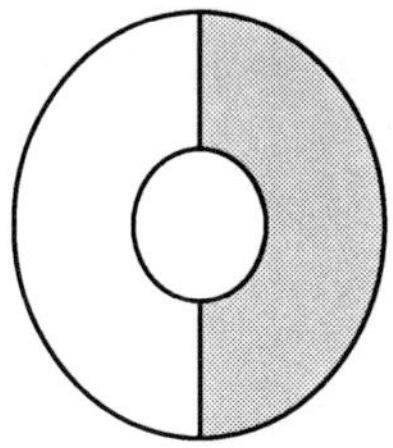

FIGURE 4.4. Bending and Axial Shear Modes $n = 1$.

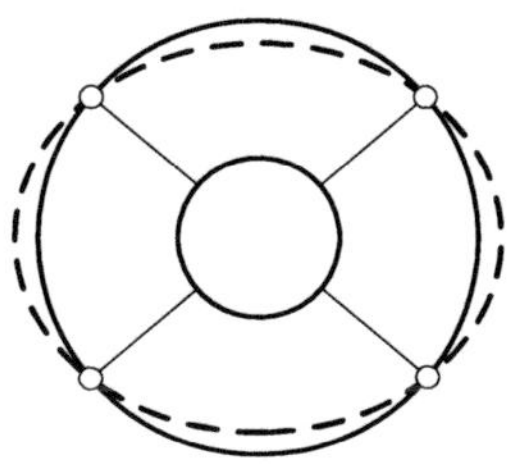

FIGURE 4.5. First Lobar Mode, $n = 2$.

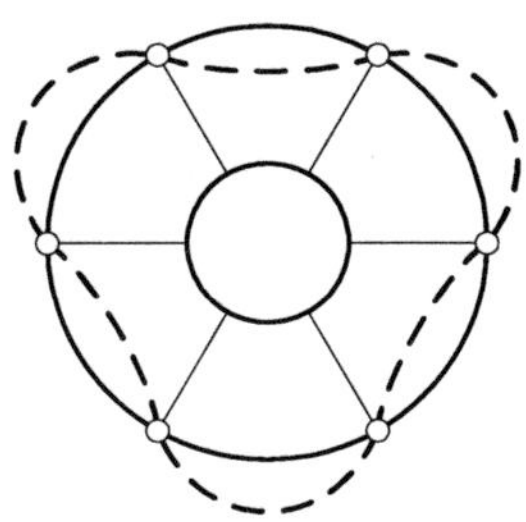

FIGURE 4.6. Second Lobar Mode, $n = 3$.

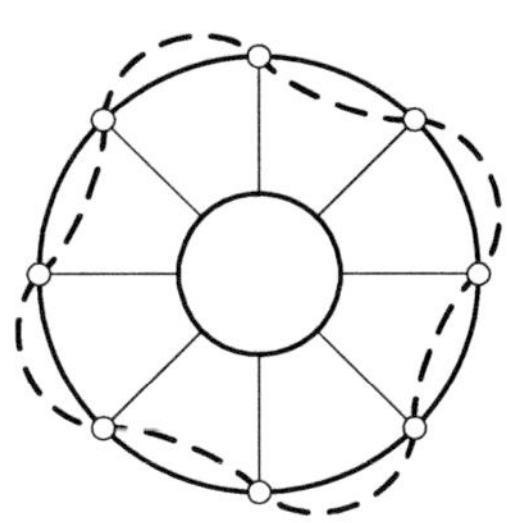

FIGURE 4.7. Third Lobar Mode, $n = 4$.

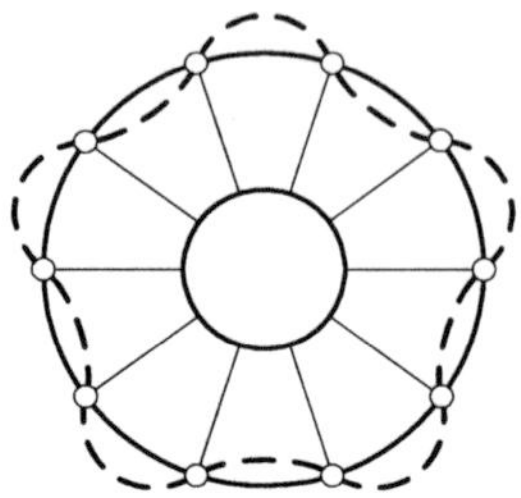

FIGURE 4.8. Fourth Lobar Mode, $n = 5$.

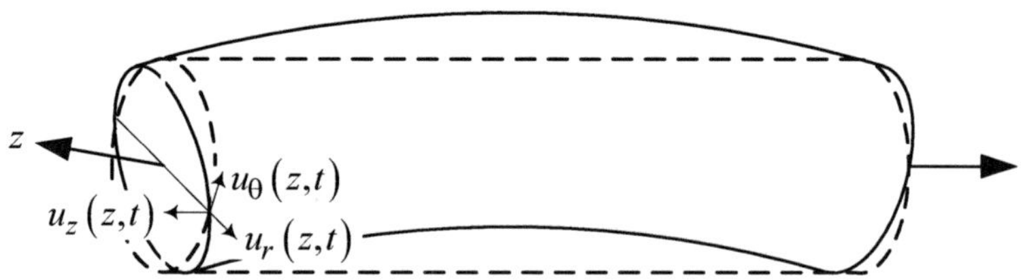

FIGURE 4.9. Rigid Body Flexural Mode, $n = 1$.

The dominance of particular modes depends upon the type of excitation. Higher modes require exciting boundary stresses with very high frequencies. Therefore, these modes are usually not noticed in most cylindrical structures. The first four lobar modes, depicted in the above figures, indicate the oscillating patterns of the cross section of the cylindrical structure. These displacement patterns are shown by the dashed lines, while the nodes are indicated by small circles.

The special cases of cylindrical vibrations include: the breathing mode $n = 0$, and the bending mode $n = 1$. The breathing mode requires radial displacement across the thickness of the cylinder in the absence of u_θ and u_z, with no nodal diameter. The bending mode is associated with no cross-sectional distortion, while bending in the axial direction occurs. Figure 4.9 depicts the bending of a cylinder for this mode.

4.5.2 Thickness Modes

Thickness modes are best introduced by using a mathematical explanation. Vibration of a cylinder is composed mathematically of an infinite number of circumferential wave numbers ($n = 0, 1, 2, 3, \ldots$), i.e. the breathing, torsional, axial, and bending modes, and the lobar modes. Furthermore, for

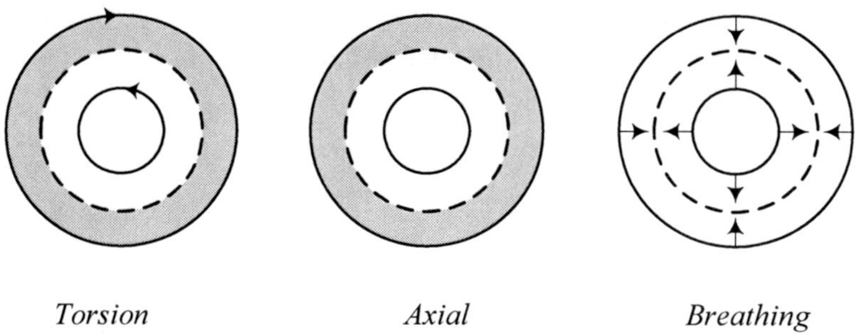

FIGURE 4.10. Modes Associated with $(n = 0)$ and One Node Circle.

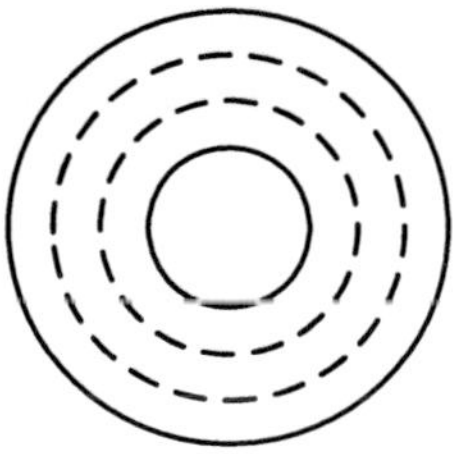

FIGURE 4.11. Modes Associated with $(n = 0)$ and Two Node Circle.

each circumferential number there is an infinite number of modes related to the infinite number of r values, where one of the displacement functions (u_r, u_θ, and u_z) vanishes. This is due to the nature of the involved Bessel functions. The thickness modes are occurring for each value of the circumferential wave number at specific values of ζ. Therefore, for the vibration of thick cylinders, there are infinite by infinite sets of vibration modes and for each mode, there are its natural frequency and its respective mode shape. Thus, whenever these types of structures are excited at a high natural frequency, a number of nodal circle(s) can occur. These nodal circle(s), describing the thickness modes, may be visualized by figures 4.10 and 4.11 that present two thickness modes for a vibrating cylindrical structure with a circumferential wave number of $n = 0$. In Figure 4.10, the first thickness mode has been presented; the dotted circle represents the thickness mode. Depending on the type of mode the associated displacement amplitude of every point on the node is zero. In other words, the dotted line indicates that on that position, no motion is observed. Furthermore, on the outside of the thickness circle nodes, the motion will be in the opposite direction to the points positioned inside the thickness circle nodes.

In Figure 4.11, the two thickness modes are presented and the two thickness node circles are shown. The points in each of the three zones for the cross section of the cylinder are moving in alternate directions. In this special case, the displacement directions could be (in, out, in) or oppositely (out, in, out). In these figures, the dashed lines indicate circular nodes.

4.5.3 Modes with Infinite Axial Wavelengths

The infinite axial wavelength is denoted by specific cases where $\zeta = 0$ $(H/L = 0)$. This special case simplifies the stress matrix equation into two problems which can be independently solved. The independent solution of the two problems is commonly referred to as uncoupling of the solutions. As shown by Armenakas et al. (1969), the $(r,\ z)$ plane motion is uncoupled from the longitudinal motion. By this uncoupling, the two problems that arise for infinite axial wavelengths are the plane strain vibration and axial (longitudinal) vibration. Plane strain vibration is denoted by radial and transverse motion, while $u_z = 0$. The axial shear vibration happens by axial motion only. The vibration observed where $\zeta = 0$ can be axial, plane strain, or a combination of both.

For the special case, Breathing mode $(n = 0)$ and $(\zeta = 0)$, the transverse shear mode is uncoupled from the motion in the plane $(r,\ z)$. Therefore, there are now three families of uncoupled solutions. This results in radial, transverse, or axial shear vibrations being observed. Also, the matrix solutions, for the plane strain and torsional modes, become identical. This results in the same natural frequencies for these modes.

4.5.4 Modes with Finite Axial Wavelengths

When ζ has a value other than zero, finite length cylinders that are simply supported can also be analyzed to determine their natural frequencies, displacement, and stress mode shapes. In this analysis, both the Bessel functions and modified Bessel functions are utilized. For $\zeta \neq 0$, the stress matrix cannot be uncoupled. Therefore, the plane strain vibration and the axial shear vibration are coupled. It should be noted that the nature of the breathing mode uncouples the transverse shear from the motion in the $(r,\ z)$ plane.

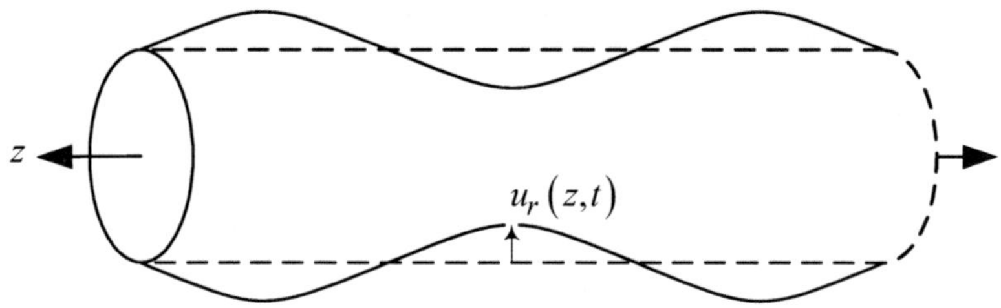

FIGURE 4.12. Radial Axisymmetric Mode.

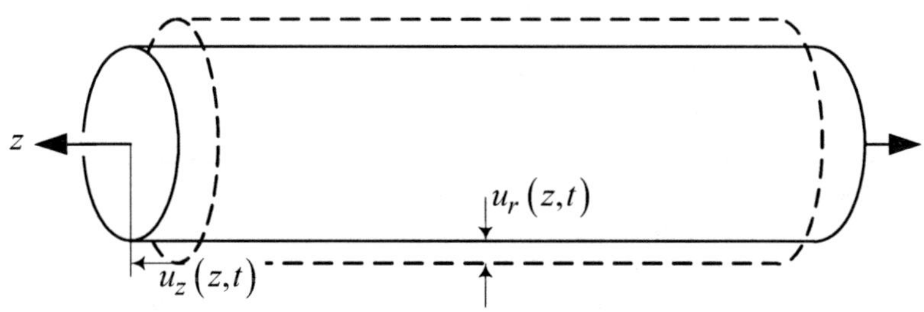

FIGURE 4.13. Axisymmetric Longitudinal and Breathing Modes.

4.5.5 Axially Symmetric Modes

The axially symmetric vibration occurs for the special case of $n = 0$, the breathing mode. As shown by Armenakas et al. (1969), the stress matrix problem can be separated into two problems and then independently solved. The two problems that arise are those of longitudinal waves and torsional waves. Longitudinal waves involve only radial and axial displacements. Torsional waves involve nodal cylinders. Figure 4.12 depicts a radial axisymmetric mode and figure 4.13 represents the longitudinal mode.

The torsional waves are more accurately described as vibration of the transverse sections of the cylinder. Each transverse section of the cylinder may rotate as a whole about the center, either clockwise or counterclockwise. The rotation direction depends upon the torsional mode. Figure 4.14 represents a typical torsional mode.

Towest torsional mode is characterized by a continued rotation in a single direction (clockwise, for example). This rotation direction is continued through the length of the axial wave. The lowest torsional mode corresponds to $\beta^2 = 0$ which was presented in equation (3.37), and hence,

$$\omega = \zeta v_2. \tag{4.9}$$

Therefore, the torsional wave problem does not involve the Bessel equations that were presented earlier.

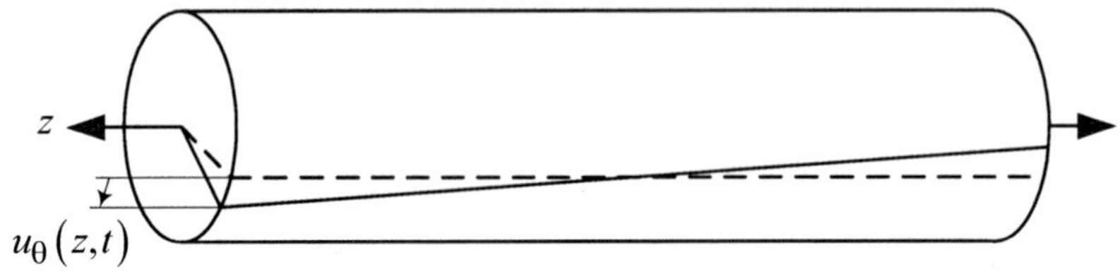

FIGURE 4.14. Axisymmetric Torsional Mode.

4.6 Forced and Free Modal Analysis

The work of Fourier illustrates that any periodic function can be represented by a series of harmonic functions. In this chapter, harmonic excitation for boundary stresses are used to calculate resonance frequencies for coupled axial, torsional, flexural, and thickness modes. These harmonic excitations are for a specific vibration mode with certain values for axial and circumferential wave numbers, and excitation frequencies. With an appropriate harmonic exciting boundary stresses, the forced vibration frequency response for a given cylinder can be determined. In order to conduct modal analysis using forced vibration, constant values for amplitudes of inner boundary stresses can be assigned.

$$\sigma_{rr} = A_1 \cos(n\theta)\cos(\omega t + \zeta z) \tag{4.10}$$

$$\tau_{r\theta} = A_2 \sin(n\theta)\cos(\omega t + \zeta z) \tag{4.11}$$

$$\tau_{rz} = A_3 \cos(n\theta)\sin(\omega t + \zeta z) \tag{4.12}$$

where:

A_k is the amplitude of the stress on the inner boundary,
n is the circumferential wave number,
ω is the frequency of excitation,
ζ is the axial wave number,
(θ, z) are coordinate directions, and
t is time.

To obtain a set of resonance frequencies for a particular axial wave length and a circumferential wave number, the matrix equation (3.129) can be used. Utilizing equations (4.10)-(4.12) for the inner boundary stresses and setting the outer boundary stresses to zero, numerical computation can be conducted to determine the stresses and displacements of the outer boundary of the vibrating cylinder for a wide range of frequencies and a small frequency interval. Then, the first six resonance frequencies for different circumferential wave numbers, n, can be computed by determining which frequency factor causes the maximum absolute value of modal displacement. This is done independently for the first six modes of vibration for a specific cylindrical geometry. For each mode, different circumferential and axial wave numbers can be considered. The computed results for the outer boundary stresses should be zero. This may be used as a verification of the developed computer program for this analysis. It should be noted that the exciting stresses may be applied to the outer boundary with no expected difference in outcomes for the resonance frequencies. The inner or outer excited boundary stresses for the cylinder must be non-zero in all three directions to avoid missing a resonance mode. Without excitation in a certain direction, resulted amplitudes for a particular mode may be too small to be identified, especially for weak modes.

To identify the resonance frequency factors, a frequency sweep can be made for the required parameters and the modal displacement data for each resonance frequency factor can be computed. To achieve higher accuracy for resonance frequency factors, groups of finer frequency factor range that include the resonant displacement, can be swept again. Thus, resonance frequency factors can be identified by determining the frequency factor that resulted in a greatest resonant displacement. Since some of the resonant frequencies cannot be identified using large frequency factor intervals, frequency sweeps should be conducted carefully. Small frequency factor intervals must be examined to assure an accurate resonance frequency factor. For example, a frequency factor interval of 0.1×10^{-5} may represent a frequency interval of 25 Hz, which is a large interval to identify the natural frequencies. An upper and lower triangular matrix method is recommended to compute the constants vector $\mathbf{X}$ in equation (3.106). The upper and lower triangular matrix method is more accurate than the matrix inversion method.

For free vibration analysis, since boundaries are free from stresses using equation (3.129), one can compute the natural frequencies. Natural frequency factors can be determined by setting the determinant of the matrix DT^{-1} equal to zero. In this section, the inner or outer boundary of the cylinder must be stressed in all three of the defined directions to avoid missing a resonating mode. Without excitation in a certain direction, resultant amplitudes of a particular mode may be too small to be numerically identifiable.

In this chapter, the first six natural frequency factors, Ω for the first five circumferential wave numbers and geometric parameters are computed and presented in a tabular format. The following values and range of parameters are considered for the computations:

Poisson's ratio: 0.3

Ratio of thickness to mean radius: $0.3 \leq H/R \leq 1.9$
Ratio of thickness to axial half wavelengths: $0 \leq H/L \leq 1.0$
Number of circumferential waves: $n = 0, 1, 2, 3, 4, 5$
Damping loss-factors: $\eta = 0.0, 0.05, 0.1, 0.2, 0.3, 0.5$, and 1.0
Normalized frequency factor: $0.003 \leq \Omega \leq 6.0$ (variable)

The computed results of natural frequency factors for a number of modes and the above-mentioned parameters are presented. The results are tabulated, comparing the natural frequency factors of elastic cylinders to those obtained by Armenakas et al. (1969) in Tables 4.1-4.25. Furthermore, the results for the first six natural frequency factors are also provided for $n = 5$ in Tables 4.26-4.30.

In summary, the results are: Tables 4.1 through 4.25 present the first six frequency factors obtained by Armenakas et. al. (1969), and those computed using the presented technique. Each table offers data for four axial wavelengths.

Tables 4.1 to 4.5 include data for $H/R = 0.3$ and $n = 0, 1, 2, 3$, and 4 .
Tables 4.6 to 4.10 provide data for $H/R = 0.4$ and $n = 0, 1, 2, 3$, and 4.
Tables 4.11 to 4.15 present data for $H/R = 0.5$ and $n = 0, 1, 2, 3$, and 4.
Tables 4.16 to 4.20 give data for $H/R = 1.0$ and $n = 0, 1, 2, 3$, and 4.
Tables 4.21 to 4.25 contain data for $H/R = 1.9$ and $n = 0, 1, 2, 3$, and 4.
Table 4.26 to 4.30 contain data for $H/R = 0.3, 0.4, 0.5, 1.0$, and 1.9, and $n = 5$.

Table 4.1 - Comparison of computed natural frequency factors with those of Ref [Armenakas et al.] for the first six modes ($H/R = 0.3$ and $n = 0$)

H/L	Resonant frequency factors, Ω						
0.0	Present	0.	0.16301	1.00303	1.87872	2.00181	3.00116
	Ref.	0.	0.16306	1.00344	1.87811	2.00174	3.00118
0.1	Present	0.13853	0.18950	1.02102	1.85702	2.03136	3.00069
	Ref.	0.13829	0.18946	1.02055	1.85681	2.03137	3.00076
0.5	Present	0.31817	0.82804	1.33441	1.74405	2.33701	3.00064
	Ref.	0.31790	0.82791	1.33441	1.74401	2.33665	3.00068
1.0	Present	0.74904	1.42011	1.91202	2.00302	2.82872	3.07883
	Ref.	0.78474	1.42002	1.91156	2.00294	2.82864	3.07880

Table 4.2 - Comparison of computed natural frequency factors with those of Ref [Armenakas et al.] for the first six modes ($H/R = 0.3$ and $n = 1$)

H/L	Resonant frequency factors, Ω						
0.0	Present	0.	0.09589	0.22528	1.00825	1.03407	1.85849
	Ref.	0.	0.09584	0.22524	1.00819	1.03403	1.85845
0.1	Present	0.06139	0.16487	0.26031	1.01832	1.04542	1.84230
	Ref.	0.06158	0.16484	0.26027	1.01826	1.04544	1.84229
0.5	Present	0.32103	0.51071	0.84139	1.13772	1.34524	1.74348
	Ref.	0.32101	0.51065	0.84138	1.13767	1.34521	1.74344
1.0	Present	0.78863	1.00487	1.41669	1.43621	1.91693	2.00881
	Ref.	0.78856	1.00483	1.41665	1.43618	1.91684	2.00872

Table 4.3 - Comparison of computed natural frequency factors with those of Ref [Armenakas et al.] for the first six modes ($H/R = 0.3$ and $n = 2$)

H/L	Resonant frequency factors, Ω						
0.0	Present	0.03611	0.19171	0.35153	1.02246	1.08220	1.82141
	Ref.	0.03596	0.19161	0.35149	1.00230	1.08213	1.82136
0.1	Present	0.05462	0.22790	0.38253	1.02959	1.09442	1.81199
	Ref.	0.05456	0.22784	0.38247	1.02966	1.09432	1.81191
0.5	Present	0.33339	0.54012	0.87988	1.15074	1.37691	1.74263
	Ref.	0.33332	0.54009	0.87975	1.15063	1.37683	1.74257
1.0	Present	0.80023	1.01920	1.41768	1.45432	1.93226	2.02644
	Ref.	0.80015	1.01911	1.41769	1.45421	1.93221	2.02639

Table 4.4 - Comparison of computed natural frequency factors with those of Ref [Armenakas et al.] for the first six modes ($H/R = 0.3$ and $n = 3$)

H/L	Resonant frequency factors, Ω						
0.0	Present	0.09559	0.28731	0.49311	1.04462	1.15526	1.78664
	Ref.	0.09556	0.28721	0.49302	1.04453	1.15517	1.78652
0.1	Present	0.10642	0.30962	0.51786	1.05175	1.16745	1.78121
	Ref.	0.10638	0.30951	0.51772	1.0517	1.16737	1.78116
0.5	Present	0.35972	0.58372	0.93781	1.17226	1.42714	1.74392
	Ref.	0.35967	0.58365	0.93779	1.1722	1.42709	1.74383
1.0	Present	0.81961	1.04229	1.42546	1.47752	1.95653	2.05682
	Ref.	0.81956	1.04226	1.42537	1.47741	1.95649	2.05677

Table 4.5 - Comparison of computed natural frequency factors with those of Ref [Armenakas et al.] for the first six modes ($H/R = 0.3$ and $n = 4$)

H/L	Resonant frequency factors, Ω						
0.0	Present	0.17002	0.38264	0.63703	1.07711	1.24642	1.75996
	Ref.	0.16997	0.38258	0.63697	1.07706	1.24635	1.75991
0.1	Present	0.17925	0.39831	0.65672	1.08285	1.25747	1.75723
	Ref.	0.17917	0.39825	0.65666	1.08277	1.25739	1.75717
0.5	Present	0.40152	0.63771	1.00891	1.20242	1.49267	1.75105
	Ref.	0.40149	0.6376	1.00885	1.20233	1.4926	1.75099
1.0	Present	0.84676	1.07351	1.43953	1.50501	1.98833	2.10052
	Ref.	0.84673	1.07349	1.43945	1.50493	1.98827	2.10048

Table 4.6 - Comparison of computed natural frequency factors with those of Ref [Armenakas et al.] for the first six modes ($H/R = 0.4$ and $n = 0$)

H/L	Resonant frequency factors, Ω						
0.0	Present	0.	0.21916	1.00618	1.88527	2.00314	3.00217
	Ref.	0.	0.21915	1.00616	1.88521	2.00315	3.00212
0.1	Present	0.15291	0.22991	1.02342	1.86262	2.03353	3.00177
	Ref.	0.15289	0.2298	1.02339	1.86255	2.0335	3.00171
0.5	Present	0.34113	0.83024	1.38841	1.74811	2.34974	3.00192
	Ref.	0.34108	0.83018	1.33836	1.74805	2.33966	3.00181
1.0	Present	0.79281	1.42481	1.91519	2.00442	2.83123	3.08013
	Ref.	0.79274	1.42474	1.91512	2.00431	2.83118	3.0801

Table 4.7 - Comparison of computed natural frequency factors with those of Ref [Armenakas et al.] for the first six modes ($H/R = 0.4$ and $n = 1$)

H/L	Resonant frequency factors, Ω						
0.0	Present	0.	0.12817	0.29693	1.01857	1.06221	1.52004
	Ref.	0.	0.12813	0.29686	1.0185	1.06217	1.52000
0.1	Present	0.05682	0.19202	0.32143	1.02542	1.07288	1.83817
	Ref.	0.05695	0.19195	0.32138	1.02538	1.07276	1.83810
0.5	Present	0.34077	0.52102	0.85321	1.15314	1.35925	1.74697
	Ref.	0.34071	0.52096	0.85315	1.15311	1.35922	1.74691
1.0	Present	0.79863	1.00901	1.41862	1.45363	1.92443	2.01521
	Ref.	0.79858	1.00896	1.41855	1.45351	1.92435	2.01516

Table 4.8 - Comparison of computed natural frequency factors with those of Ref [Armenakas et al.] for the first six modes ($H/R = 0.4$ and $n = 2$)

H/L	Resonant frequency factors, Ω						
0.0	Present	0.06216	0.25613	0.45802	1.04061	1.14901	1.80170
	Ref.	0.06211	0.25593	0.45793	1.04057	1.14894	1.80159
0.1	Present	0.07467	0.28584	0.48176	1.04804	1.16051	1.79503
	Ref.	0.07456	0.28573	0.48162	1.04801	1.16046	1.79501
0.5	Present	0.35448	0.57311	0.91569	1.17673	1.41891	1.74636
	Ref.	0.35445	0.57309	0.91566	1.17671	1.41882	1.74632
1.0	Present	0.81671	1.03499	1.42068	1.48476	1.95017	2.04921
	Ref.	0.81660	1.03494	1.42072	1.48478	1.95012	2.04913

Table 4.9 - Comparison of computed natural frequency factors with those of Ref [Armenakas et al.] for the first six modes ($H/R = 0.4$ and $n = 3$)

H/L	Resonant frequency factors, Ω						
0.0	Present	0.14940	0.38311	0.63719	1.08285	1.27361	1.76325
	Ref.	0.14935	0.38306	0.63714	1.08277	1.27357	1.76321
0.1	Present	0.16784	0.40070	0.65613	1.08863	1.28401	1.76050
	Ref.	0.16774	0.40061	0.65601	1.08854	1.28391	1.76045
0.5	Present	0.39801	0.64377	1.00345	1.21633	1.50885	1.75444
	Ref.	0.39794	0.64374	1.00342	1.2163	1.5088	1.75441
1.0	Present	0.84784	1.07577	1.43511	1.52371	1.98840	2.10849
	Ref.	0.84781	1.07579	1.4351	1.52367	1.98837	2.10841

Table 4.10 - Comparison of computed natural frequency factors with those of Ref [Armenakas et al.] for the first six modes ($H/R = 0.4$ and $n = 4$)

H/L	Resonant Frequency Factors, Ω						
0.0	Present	0.27422	0.50930	0.81564	1.13851	1.41924	1.74703
	Ref.	0.27416	0.50922	0.81561	1.13847	1.41926	1.74701
0.1	Present	0.28137	0.52139	0.83026	1.14421	1.42801	1.74687
	Ref.	0.28134	0.52136	0.83022	1.14419	1.42795	1.74684
0.5	Present	0.49699	0.72693	1.10202	1.27141	1.61328	1.78673
	Ref.	0.49698	0.72697	1.102	1.27137	1.61323	1.78678
1.0	Present	0.89261	1.12901	1.46097	1.56882	2.03661	2.19128
	Ref.	0.89257	1.12897	1.46091	1.56878	2.03657	2.19128

Table 4.11 - Comparison of computed natural frequency factors with those of Ref [Armenakas et al.] for the first six modes ($H/R = 0.5$ and $n = 0$)

H/L	Resonant Frequency Factors, Ω						
0.0	Present	0.	0.2670	1.00971	1.89411	2.00503	3.00335
	Ref.	0.	2.7673	1.00969	1.89408	2.00501	3.00336
0.1	Present	0.15702	0.28239	1.02704	1.87036	2.03641	3.00301
	Ref.	0.15695	0.28235	1.02701	1.87032	2.03642	3.00297
0.5	Present	0.36921	0.83342	1.34367	1.75357	2.34739	3.00331
	Ref.	0.36919	0.8334	1.34362	1.75354	2.34734	3.00328
1.0	Present	0.80320	1.43102	1.92001	2.00619	2.83461	3.08177
	Ref.	0.80317	1.4310	1.91993	2.00612	2.83465	3.08180

Table 4.12 - Comparison of computed natural frequency factors with those of Ref [Armenakas et al.] for the first six modes ($H/R = 0.5$ and $n = 1$)

H/L	Resonant Frequency Factors, Ω						
0.0	Present	0.	0.16070	0.36514	1.02384	1.10033	1.84605
	Ref.	0.	0.16069	0.36513	1.02382	1.10034	1.84602
0.1	Present	0.0523	0.21866	0.38415	1.03447	1.11035	1.83411
	Ref.	0.05213	0.21862	0.38412	1.03443	1.11037	1.83414
0.5	Present	0.35888	0.53635	0.86721	1.17273	1.37983	1.75152
	Ref.	0.35884	0.53632	0.86723	1.17273	1.3798	1.75153
1.0	Present	0.81023	1.01476	1.42106	1.47601	1.93400	2.02428
	Ref.	0.81027	1.01473	1.42103	1.47600	1.93396	2.02431

Table 4.13 - Comparison of computed natural frequency factors with those of Ref [Armenakas et al.] for the first six modes ($H/R = 0.5$ and $n = 2$)

H/L	Resonant Frequency Factors, Ω						
0.0	Present	0.09377	0.32031	0.55468	1.06539	1.23762	1.78518
	Ref.	0.09383	0.32034	0.55464	1.06533	1.23759	1.78511
0.1	Present	0.10292	0.34561	0.57411	1.07268	1.24787	1.78072
	Ref.	0.10289	0.34556	0.57411	1.07271	1.24783	1.78065
0.5	Present	0.37305	0.61402	0.95429	1.21001	1.47873	1.75287
	Ref.	0.37308	0.6140	0.95428	1.21002	1.47867	1.75288
1.0	Present	0.83409	1.05605	1.42503	1.52328	1.97016	2.08287
	Ref.	0.83411	1.05603	1.42505	1.52322	1.97017	2.08290

Table 4.14 - Comparison of computed natural frequency factors with those of Ref [Armenakas et al.] for the first six modes ($H/R = 0.5$ and $n = 3$)

H/L	Resonant Frequency Factors, Ω						
0.0	Present	0.23454	0.47801	0.76405	1.13182	1.42104	1.75286
	Ref.	0.23251	0.47797	0.76409	1.13183	1.42106	1.75283
0.1	Present	0.23919	0.49257	0.77935	1.13805	1.42955	1.75207
	Ref.	0.23916	0.49251	0.77933	1.13803	1.42952	1.75210
0.5	Present	0.44122	0.71208	1.06774	1.27300	1.61079	1.78285
	Ref.	0.44121	0.71206	1.06769	1.27295	1.61076	1.78280
1.0	Present	0.87869	1.11808	1.44905	1.58022	2.02118	2.18172
	Ref.	0.87864	1.11805	1.44901	1.58019	2.02114	2.18164

Table 4.15 - Comparison of computed natural frequency factors with those of Ref [Armenakas et al.] for the first six modes ($H/R = 0.5$ and $n = 4$)

H/L	Resonant Frequency Factors, Ω						
0.0	Present	0.38839	0.63279	0.96802	1.22005	1.59996	1.78585
	Ref.	0.38836	0.63276	0.96798	1.22007	1.59990	1.78581
0.1	Present	0.39401	0.64269	0.97918	1.22579	1.60522	1.78901
	Ref.	0.39397	0.64279	0.97921	1.22574	1.60519	1.78891
0.5	Present	0.55009	0.82386	1.18787	1.35824	1.71428	1.89824
	Ref.	0.55006	0.82384	1.18781	1.35817	1.71424	1.89820
1.0	Present	0.94455	1.19235	1.49093	1.64638	2.08588	2.29947
	Ref.	0.94460	1.19236	1.49090	1.64634	2.08582	2.29949

Table 4.16 - Comparison of computed natural frequency factors with those of Ref [Armenakas et al.] for the first six modes ($H/R = 1.0$ and $n = 0$)

H/L	Resonant Frequency Factors, Ω						
0.0	Present	0.	0.60063	1.04126	1.99844	2.02375	3.01637
	Ref.	0.	0.60059	1.04127	1.99847	2.02371	3.01636
0.1	Present	0.16032	0.59785	1.06123	1.95261	2.07682	3.01613
	Ref.	0.16029	0.59789	1.0612	1.95264	2.07688	3.01611
0.5	Present	0.55372	0.87516	1.39569	1.81700	2.39292	3.01849
	Ref.	0.55369	0.87513	1.39566	1.81691	2.39297	3.01856
1.0	Present	0.89203	1.49805	1.97702	2.02383	2.87638	3.09891
	Ref.	0.89200	1.49809	1.97696	2.02388	2.87635	3.09887

Table 4.17 - Comparison of computed natural frequency factors with those of Ref [Armenakas et al.] for the first six modes ($H/R = 1.0$ and $n = 1$)

H/L	Resonant Frequency Factors, Ω						
0.0	Present	0.	0.32700	0.62737	1.11902	1.46298	1.83625
	Ref.	0.	0.32698	0.62732	1.11903	1.46294	1.83627
0.1	Present	0.03567	0.36447	0.63778	1.12802	1.47032	1.83142
	Ref.	0.03563	0.36452	0.63771	1.12806	1.47034	1.83143
0.5	Present	0.39268	0.67211	0.94887	1.30141	1.64581	1.79861
	Ref.	0.39265	0.67212	0.94885	1.30142	1.64485	1.79854
1.0	Present	0.86593	1.07650	1.44301	1.65805	2.00044	2.14271
	Ref.	0.86589	1.07651	1.44292	1.65803	2.00047	2.14177

Table 4.18 - Comparison of computed natural frequency factors with those of Ref [Armenakas et al.] for the first six modes ($H/R = 1.0$ and $n = 2$)

H/L	Resonant Frequency Factors, Ω						
0.0	Present	0.31061	0.62227	0.89218	1.33056	1.72918	1.97099
	Ref.	0.31064	0.62230	0.89214	1.33060	1.72920	1.97090
0.1	Present	0.31303	0.63767	0.90340	1.33591	1.73081	1.97507
	Ref.	0.31299	0.63760	0.90339	1.33586	1.73082	1.97500
0.5	Present	0.47665	0.84182	1.13577	1.46501	1.78009	2.07508
	Ref.	0.47661	0.84179	1.13570	1.46504	1.78015	2.07516
1.0	Present	0.91162	1.21529	1.48147	1.81264	2.06432	2.35507
	Ref.	0.91153	1.21522	1.48146	1.81254	2.06424	2.35509

Table 4.19 - Comparison of computed natural frequency factors with those of Ref [Armenakas et al.] for the first six modes ($H/R = 1.0$ and $n = 3$)

H/L	Resonant Frequency Factors, Ω						
0.0	Present	0.65844	0.88370	1.21732	1.61846	1.79892	2.35097
	Ref.	0.65849	0.88365	1.21733	1.61842	1.79897	2.35095
0.1	Present	0.65951	0.89336	1.22517	1.62191	1.80255	2.34680
	Ref.	0.65957	0.89336	1.22510	1.62188	1.80258	2.34683
0.5	Present	0.74093	1.05807	1.37811	1.72001	1.8806	2.37606
	Ref.	0.74090	1.05812	1.37817	1.72004	1.8813	2.37614
1.0	Present	1.06027	1.37801	1.62387	2.01921	2.19132	2.55683
	Ref.	1.06030	1.37805	1.62382	2.01918	2.19125	2.55677

Table 4.20 - Comparison of computed natural frequency factors with those of Ref [Armenakas et al.] for the first six modes ($H/R = 1.0$ and $n = 4$)

H/L	Resonant Frequency Factors, Ω						
0.0	Present	0.95891	1.12625	1.55542	1.92250	2.00328	2.58983
	Ref.	0.95897	1.12632	1.55537	1.92242	2.00330	2.58987
0.1	Present	0.95943	1.13382	1.56062	1.92501	2.00752	2.59121
	Ref.	0.95948	1.13385	1.56058	1.92503	2.00757	2.59128
0.5	Present	1.00967	1.27070	1.66533	2.00609	2.09971	2.63101
	Ref.	1.00961	1.27067	1.66535	2.00616	2.09966	2.63097
1.0	Present	1.24733	1.55409	1.86718	2.25758	2.38021	2.79232
	Ref.	1.24734	1.55401	1.86716	2.25763	2.38029	2.79234

Table 4.21 - Comparison of computed natural frequency factors with those of Ref [Armenakas et al.] for the first six modes ($H/R = 1.9$ and $n = 0$)

H/L	Resonant Frequency Factors, Ω						
0.0	Present	0.	1.19076	1.22962	2.18359	3.11852	3.17102
	Ref.	0.	1.19078	1.22956	2.18355	3.11847	3.17090
0.1	Present	0.16048	1.15491	1.27831	2.18570	3.11781	3.17862
	Ref.	0.16086	1.15497	1.27838	2.18575	3.11777	3.17854
0.5	Present	0.73574	1.15279	1.62351	2.24689	3.13850	3.32038
	Ref.	0.73570	1.15285	1.62355	2.24687	3.13854	3.32032
1.0	Present	1.09661	1.73153	2.13091	2.52748	3.25918	3.66571
	Ref.	1.09668	1.73159	2.13088	2.52757	3.25915	3.66578

Table 4.22 - Comparison of computed natural frequency factors with those of Ref [Armenakas et al.] for the first six modes ($H/R = 1.9$ and $n = 1$)

H/L	Resonant Frequency Factors, Ω						
0.0	Present	0.	0.57201	0.84144	1.64887	1.99653	2.20241
	Ref.	0.	0.57205	0.8415	1.64883	1.99648	2.20236
0.1	Present	0.02490	0.59572	0.88173	1.64911	2.00181	2.20791
	Ref.	0.02482	0.59561	0.88167	1.64920	2.00173	2.20786
0.5	Present	0.37001	0.90074	1.09051	1.66998	2.09682	2.33612
	Ref.	0.36991	0.90070	1.09049	1.68995	2.09668	2.33616
1.0	Present	0.87957	1.24771	1.54552	2.01172	2.29524	2.62975
	Ref.	0.87963	1.24776	1.54558	2.01168	2.29528	2.62982

Table 4.23 - Comparison of computed natural frequency factors with those of Ref [Armenakas et al.] for the first six modes ($H/R = 1.9$ and $n = 2$)

H/L	Resonant Frequency Factors, Ω						
0.0	Present	0.72557	0.94735	1.35982	2.07981	2.42737	2.93408
	Ref.	0.72554	0.94727	1.35870	2.07987	2.42738	2.93410
0.1	Present	0.72182	0.96209	1.36564	2.08146	2.43034	2.93682
	Ref.	0.72186	0.96213	1.36560	2.08141	2.43031	2.93680
0.5	Present	0.75746	1.17249	1.53017	2.12839	2.49581	2.99262
	Ref.	0.75748	1.17258	1.53016	2.12836	2.49578	2.99255
1.0	Present	1.05153	1.48609	1.90947	2.36371	2.67252	3.13579
	Ref.	1.05158	1.48616	1.90952	2.36363	2.67241	3.13575

Table 4.24 - Comparison of computed natural frequency factors with those of Ref [Armenakas et al.] for the first six modes ($H/R = 1.9$ and $n = 3$)

H/L	Resonant Frequency Factors, Ω						
0.0	Present	1.12101	1.30305	1.86544	2.48597	2.87071	3.51898
	Ref.	1.121040	1.30300	1.86541	2.48593	2.87063	3.51893
0.1	Present	1.11894	1.31229	1.87102	2.48778	2.87294	3.51701
	Ref.	1.11891	1.31223	1.87098	2.48782	2.87287	3.51697
0.5	Present	1.13093	1.46463	1.99818	2.53759	2.92530	3.51529
	Ref.	1.13098	1.46469	1.99814	2.53761	2.92534	3.51536
1.0	Present	1.31703	1.74033	2.30136	2.73571	3.08227	3.59928
	Ref.	1.31705	1.74040	2.30141	2.73575	3.08235	3.59928

Table 4.25 - Comparison of computed natural frequency factors with those of Ref [Armenakas et al.] for the first six modes ($H/R = 1.9$ and $n = 4$)

H/L	Resonant Frequency Factors, Ω						
0.0	Present	1.46231	1.64918	2.36764	2.87895	3.28911	3.93326
	Ref.	1.46237	1.64924	2.36762	2.87892	3.28904	3.93327
0.1	Present	1.46269	1.65530	2.37159	2.88091	3.29106	3.93381
	Ref.	1.46266	1.65521	2.37156	2.88085	3.29102	3.93377
0.5	Present	1.47591	1.76911	2.46815	2.92974	3.33792	3.94971
	Ref.	1.47586	1.76909	2.46817	2.92969	3.33795	3.94973
1.0	Present	1.61059	2.00871	2.70408	3.10625	3.48378	4.02779
	Ref.	1.61050	2.00866	2.70406	3.10630	3.48373	4.02776

Table 4.26 - Frequency factor for the fourth lobar mode
($H/R = 0.3$ and $n = 5$)

H/L	Resonant Frequency Factors, Ω						
0.0	Present	0.25417	0.47832	0.77942	1.11734	1.35032	1.74527
0.1	Present	0.26239	0.49354	0.79532	1.12208	1.35948	1.74486
0.5	Present	0.45731	0.69919	1.08648	1.24136	1.56843	1.76942
1.0	Present	0.59445	0.88231	1.11230	1.45931	1.53638	2.02629

Table 4.27 - Frequency factor for the fourth lobar mode
($H/R = 0.4$ and $n = 5$)

H/L	Resonant Frequency Factors, Ω						
0.0	Present	0.39845	0.63436	0.98639	1.20752	1.56438	1.77158
0.1	Present	0.40439	0.64342	0.99721	1.21343	1.57142	1.77327
0.5	Present	0.56131	0.81926	1.20106	1.34062	1.70538	1.87021
1.0	Present	0.95431	1.19153	1.49728	1.62047	2.09436	2.28832

Table 4.28 - Frequency factor for the fourth lobar mode
($H/R = 0.5$ and $n = 5$)

H/L	Resonant Frequency Factors, Ω						
0.0	Present	0.55368	0.78421	1.15631	1.32741	1.69733	1.95373
0.1	Present	0.55521	0.79153	1.16317	1.33211	1.70053	1.95942
0.5	Present	0.68023	0.94432	1.30542	1.45935	1.79428	2.08726
1.0	Present	1.03348	1.28447	1.54826	1.72306	2.16430	2.39241

Table 4.29 - Frequency factor for the fourth lobar mode
($H/R = 1.0$ and $n = 5$)

H/L	Resonant Frequency Factors, Ω						
0.0	Present	1.20815	1.36134	1.90436	2.21137	2.30341	2.87831
0.1	Present	1.20908	1.36727	1.90722	2.21417	2.30722	2.87902
0.5	Present	1.24827	1.48135	1.98841	2.28834	2.38420	2.91716
1.0	Present	1.43739	1.73420	2.16539	2.50414	2.62235	3.06425

Table 4.30 - Frequency factor for the fourth lobar mode
($H/R = 1.9$ and $n = 5$)

H/L	Resonant Frequency Factors, Ω						
0.0	Present	1.78452	2.11367	2.85564	3.25742	3.70863	4.31273
0.1	Present	1.78471	1.99432	2.85932	3.26486	3.71034	4.33920
0.5	Present	1.80234	2.08214	2.93135	3.31219	3.75421	4.35981
1.0	Present	1.93018	2.28934	3.11183	3.47362	3.89223	4.43732

Table 4.31 presents a comparison of computed natural frequency factors for the first three torsional modes ($n = 0$) for different values of H/L at $H/R = 0.3$ and their comparisons with those of Armenakas et al. (1963).

Table 4.31 - Comparison of computed torsional resonant frequency factors for different H/L with those of Ref [Armenakas et al.]
($H/R = 0.3$)

H/L	Resonant Frequency Factors, Ω			
0.0	Present	0.0	1.0172	2.0088
	Ref.	0.0	1.0172	2.0089
0.1	Present	0.1000	1.0220	2.0112
	Ref.	0.1000	1.0220	2.0110
0.2	Present	0.2000	1.0368	2.0188
	Ref.	0.2000	1.0270	2.0190
0.4	Present	0.4000	1.0930	2.0480
	Ref.	0.4000	1.0930	2.0480
0.6	Present	0.6000	1.1809	2.0964
	Ref.	0.6000	1.1810	2.0960
0.8	Present	0.8000	1.2940	2.1622
	Ref.	0.8000	1.2940	2.1620
1.0	Present	1.0000	1.4264	2.2438
	Ref.	1.0000	1.4264	2.2440

In addition, the effects of the material damping on the resonant frequency factors and modal loss factors associated with different circumferential wave numbers at different thickness to half wavelength ratios for visco-elastic thick cylinders are presented. The modal parameters for thick visco-elastic cylinders are achieved by using the radial, transverse, and axial displacement responses for forced vibrations when the cylinder is subjected to boundary stresses. These responses can be computed for any point on the medium. These frequency responses can be constructed in the form of Nyquist plots for a wide range of frequencies. Using an adequate number of data points around the vicinity of the resonance and assuming a single degree of freedom, the circle-fit method is employed to determine frequency, amplitude, and the modal loss factor at each resonance. The circle-fit method, which is described by Ewins (1978), maps the frequency response data points around a circle in the Nyquist diagram. It should be noted that the accuracy of these results will increase if the contributions of the nearby mode(s) does/do not change the response significantly for the data points near the resonance under consideration.

Table 4.32 presents the extracted resonant frequency factors for a cylinder with $H/R = 0.3$, for the circumferential wave numbers of $n = 2$, and for three different values of material loss factors. This table shows that, as expected, there is no significant difference in resonant frequency factors for different levels of material damping. Although, in a few cases, the resonant frequency factors have slightly changed, this is due to the fact that the responses for these resonances have no significant peaks.

Table 4.32 - Resonant frequency factors for different material loss factors (η) and thickness to half wavelength numbers ($H/R = 0.3$ and $n = 2$).

		Resonant Frequency Factors, Ω				
H/L	η	1	2	3	4	5
0.0	0.01	0.0359	0.1919	0.3514	1.0222	1.0822
	0.02	0.0359	0.1927	0.3514	1.0222	1.0820
	0.05	0.0359	0.1963	0.3514	1.0216	1.0812
0.1	0.01	0.0499	0.2279	0.3825	1.0294	1.0949
	0.02	0.0499	0.2283	0.3825	1.0282	1.0947
	0.05	0.0501	0.2303	0.3825	1.0250	1.0925
0.2	0.01	0.1069	0.2933	0.4721	1.0463	1.1343
	0.02	0.1069	0.2933	0.4721	1.0441	1.1341
	0.05	0.1069	0.2945	0.4721	1.0453	1.1331
0.4	0.01	0.2513	0.4507	0.7323	1.1090	1.2810
	0.02	0.2513	0.4507	0.7321	1.1078	1.2808
	0.05	0.2569	0.4507	0.7321	1.1044	1.2796
0.6	0.01	0.4203	0.6329	1.0182	1.1989	1.4815
	0.02	0.4219	0.6329	1.0177	1.1983	1.4815
	0.05	0.4251	0.6341	1.0159	1.1947	1.4811
0.8	0.01	0.6061	0.8243	1.2548	1.3166	1.7051
	0.02	0.6083	0.8245	1.2534	1.3158	1.7055
	0.05	0.6107	0.8249	1.2558	1.3122	1.7079
1.0	0.01	0.7951	1.0192	1.4224	1.4543	1.9327
	0.02	0.8031	1.0196	1.4224	1.4541	1.9345
	0.05	0.8047	1.0184	1.4224	1.4527	1.9367

Tables 4.33 through 4.38 present modal loss factors for different values of material loss factors of $\eta = 0.01, 0.02$, and 0.05. The considered ratios of thickness to half wavelength: $H/L = 0.0, 0.1, 0.2, 0.4, 0.6, 0.8$, and 1.0, and the circumferential wave numbers: $n = 0, 1, 2, 3$, and 4 are presented. The special case of torsional vibration with $n = 0$ (axisymmetric) is presented in table 4.37. It should be noted that ★ indicates that data is not available, and ★★ indicates that data was not reliable due to the damped resonant peaks. The modal loss factors presented in these tables indicate that an

increasing value of modal damping requires higher material loss factor. This relationship is evident in the presented tables.

Table 4.33 - Resonant frequency factors for different material loss factors (η) and thickness to half wavelength numbers ($H/R = 0.3$ and $n = 2$).

		Resonant Frequency Factors, Ω				
H/L	η	1	2	3	4	5
0.0	0.01	★	1.634e-3	1.023e-2	1. 912e-2	1.939e-2
	0.02	★	3.262e-3	2.045e-2	3.698e-2	2.002e-2
	0.05	★	8.152e-3	4.687e-2	7.434e-2	9.920e-2
0.1	0.01	3.650e-4	2.254e-3	1.016e-2	1.910e-2	2.651e-2
	0.02	3.483e-4	4.502e-3	2.042e-3	3.663e-2	4.043e-2
	0.05	4.914e-4	1.124e-2	4.401e-2	★★	1. 003e-1
0.2	0.01	5.132e-3	2.988e-3	1. 002e-2	1. 972e-2	2.018e-2
	0.02	1. 042e-2	5.944e-3	2.006e-2	3.766e-2	3.932e-2
	0.05	2.972e-2	1.456e-2	7.118e-2	1. 084e-1	1.060e-1
0.4	0.01	1.118e-2	4.852e-3	1.014e-2	2.586e-2	1. 973e-2
	0.02	2.230e-2	9.726e-3	2.045e-2	4.507e-2	4.100e-2
	0.05	5.565e-2	2.463e-2	5.026e-2	4.796e-2	6.690e-2
0.6	0.01	1.34ge-2	6.924e-3	1.100e-2	2.427e-2	2.170e-2
	0.02	2.683e-2	1.383e-2	2.196e-2	5.341e-2	4.251e-2
	0.05	6.280e-2	3.476e-2	5.707e-2	1.078e-1	7.522e-2
0.8	0.01	2.912e-4	9.633e-3	1.241e-2	2.430e-2	2.343e-2
	0.02	6.294e-3	1. 937e-2	2.576e-2	5.616e-2	4.326e-2
	0.05	1.733e-3	5 .165e-2	5.972e-2	1.095e-1	4.461e-1
1.0	0.01	7.440e-3	1.455e-2	1.423e-2	2.361e-2	2.540e-2
	0.02	1.01le-1	2.933e-2	2.924e-2	4.391e-2	4.345e-2
	0.05	★★	6.913e-2	7.850e-2	5.322e-1	1.410e-1

Table 4.34 - Resonant frequency factors for different material loss factors (η) and thickness to half wavelength numbers ($H/R = 0.3$ and $n = 2$).

		Resonant Frequency Factors, Ω				
H/L	η	1	2	3	4	5
0.0	0.01	★	9.737e-3	2.253e-3	1.046e-2	1.034e-2
	0.02	★	1.583e-2	4.505e-3	2.113e-2	2.062e-2
	0.05	★	3.073e-2	1.133e-2	8.550e-2	6.423e-2
0.1	0.01	1. 073e-2	4.776e-3	2.687e-3	1.060e-2	1.047e-2
	0.02	2.413e-2	9.747e-3	5.360e-3	2.023e-2	2.141e-2
	0.05	9.763e-2	2.795e-2	1.338e-2	1.918e-1	3.826e-2
0.2	0.01	1.422e-2	1.908e-3	3.351e-3	1.063e-2	1. 033e-2
	0.02	2.927e-2	3.795e-3	6.703e-3	2 .163e-2	2.096e-2
	0.05	2.390e-2	9.286e-3	1. 675e-2	★★	4.573e-2
0.4	0.01	1.623e-2	7 .353e-3	6.184e-3	1. 097e-2	1. 051e-2
	0.02	3.268e-2	1.242e-2	1.232e-2	2.186e-2	2.045e-2
	0.05	6.862e-2	7.306e-3	2.911e-2	5.478e-2	4.504e-2
0.6	0.01	1. 664e-2	8.661e-3	7.640e-3	1.185e-2	1.127e-2
	0.02	3.274e-2	7.571e-4	1.548e-2	2.420e-2	2.236e-2
	0.05	8.814e-2	1.405e-2	3.760e-2	5.800e-2	5.881e-2
0.8	0.01	1. 678e-2	1.218e-3	1.007e-2	1.295e-2	1.243e-2
	0.02	3.458e-2	8.352e-3	2.013e-2	2.598e-2	2.530e-2
	0.05	8.766e-2	3.476e-2	4.593e-2	6.321e-2	6.534e-2
1.0	0.01	5.081e-2	6.050e-3	6.837e-2	1.466e-2	1.472e-2
	0.02	4.663e-2	1.693e-2	6.250e-2	3.826e-2	2.998e-2
	0.05	8.718e-2	4.783e-2	★★	6.941e-2	9.415e-2

Table 4.35 - Resonant frequency factors for different material loss factors (η) and thickness to half wavelength numbers ($H/R = 0.3$ and $n = 2$).

		Resonant Frequency Factors, Ω				
H/L	η	1	2	3	4	5
0.0	0.01	3.853e-4	3.742e-3	3.527e-3	1.068e-2	1.083e-2
	0.02	7.402e-4	7.107e-3	7.025e-3	2.138e-2	2.195e-2
	0.05	1.807e-3	1.310e-2	1.753e-2	5.333e-2	5.231e-2
0.1	0.01	1.529e-1	1.208e-3	3.784e-3	1.057e-2	1.088e-2
	0.02	1.330e-1	2.291e-3	7.576e-3	2.076e-2	2.147e-2
	0.05	6.505e-2	4.193e-3	1.884e-2	5.314e-2	4.817e-2
0.2	0.01	6.557e-2	3.217e-4	4.418e-3	1.208e-2	1.095e-2
	0.02	6.756e-2	6.072e-4	8.833e-3	2.197e-2	2.171e-2
	0.05	7.841e-2	1.630e-3	2.231e-2	5.204e-2	5.525e-2
0.4	0.01	3.012e-2	4.602e-3	1.678e-2	1.116e-2	1.114e-2
	0.02	6.312e-2	9.184e-3	2.693e-2	2.238e-2	2.262e-2
	0.05	4.535e-1	2.229e-2	5.340e-2	6.284e-2	5.651e-2
0.6	0.01	2.624e-2	7.007e-3	1.527e-2	1.192e-2	1.185e-2
	0.02	5.451e-2	1.398e-2	2.013e-2	2.317e-2	2.396e-2
	0.05	1.671e-1	3.408e-2	6.542e-2	6.055e-2	6.113e-2
0.8	0.01	2.558e-2	8.985e-3	1.136e-2	1.308e-2	1.312e-2
	0.02	5.009e-2	1.787e-2	2.216e-2	2.748e-2	2.635e-2
	0.05	1.134e-1	5.079e-2	4.845e-2	6.662e-2	7.436e-2
1.0	0.01	4.775e-2	1.081e-2	1.802e-1	1.479e-2	1.510e-2
	0.02	9.330e-2	2.163e-2	2.370e-1	2.912e-2	3.082e-2
	0.05	7.209e-1	5.847e-2	5.673e-1	6.448e-2	7.357e-2

Table 4.36 - Resonant frequency factors for different material loss factors (η) and thickness to half wavelength numbers ($H/R = 0.3$ and $n = 2$).

		Resonant Frequency Factors, Ω				
H/L	η	1	2	3	4	5
	0.01	9.705e-4	1.236e-3	4.942e-3	1.092e-2	1.157e-2
0.0	0.02	1.919e-3	2 .343e-3	9.868e-3	2.183e-2	3.055e-2
	0.05	4.772e-3	4.093e-3	2.487e-2	5.476e-2	5.930e-2
	0.01	7.586e-3	2.166e-4	5.131e-3	1.092e-2	1.164e-2
0.1	0.02	6.719e-l	3 .975e-3	1. 025e-2	2.116e-2	2.291e-2
	0.05	5.552e-2	2.107e-4	2.498e-2	5.509e-2	6.120e-2
	0.01	2.863e-1	1. 152e-2	5.655e-3	1.103e-2	1.175e-2
0.2	0.02	5.285e-2	1.944e-2	1.132e-2	2.245e-2	3.131e-2
	0.05	5.281e-l	3.835e-2	3.679e-2	5.343e-2	5.950e-2
	0.01	8.292e-2	3.798e-3	7.161e-]	1.172e-2	1.199e-2
0.4	0.02	★★	7.588e-3	1.429e-2	2.419e-2	3.245e-2
	0.05	3.052e-l	1.910e-2	3.559e-2	★★	5.931e-2
	0.01	5.578e-2	6.374e-3	1.743e-2	1.199e•2	1.269e-2
0.6	0.02	1.353e-l	1.275e•2	3.161e-2	2.449e-2	2.637e-2
	0.05	★★	3.092e-2	6.501e-2	5.935e-2	5.975e-2
	0.01	4.383e-2	8.494e-]	1.345e-2	1.371e-2	1.378e-2
0.8	0.02	1.109e-l	2.245e-2	2.647e•2	2.794e-2	2.682e-2
	0.05	1.730e-l	4.172e-2	★★	7.359e-2	7 .677e-2
	0.01	3.532e-2	1.053e-2	★★	1.519e-2	1.531e-2
1.0	0.02	8.097e-2	2.076e-2	1.494e-l	2.986e-2	3.291e-2
	0.05	1.275e-l	5.12]e-2	★★	7.349e-2	6.712e-2

Table 4.37 - Resonant frequency factors for different material loss factors (η) and thickness to half wavelength numbers ($H/R = 0.3$ and $n = 2$).

		Resonant Frequency Factors, Ω				
H/L	η	1	2	3	4	5
0.0	0.01	1.713e-3	★	6.374e-3	1. 124e-2	1.242e-2
	0.02	3.406e-3	★	1.281e-2	2.231e-2	2.468e-2
	0.05	8.518e-3	3.410e-3	3.233e-2	5.614e-2	6.348e-2
0.1	0.01	2.268e-1	★★	6.531e-2	1.116e-2	1.249e-2
	0.02	5.434e-2	2.017e-3	1.302e-2	2.190e-2	2.533e-2
	0.05	6.002e-2	5.733e-3	3.183e-2	5.753e-2	5.910e-2
0.2	0.01	4.979e-2	2 .075e-3	6.955e-3	1.120e-2	1.270e-2
	0.02	3.120e-2	4.168e-3	1.388e-2	2.262e-2	2.521e-2
	0.05	2.368e-1	1.085e-2	3.146e-2	6.742e-2	6.446e-2
0.4	0.01	1.130e-1	3.372e-3	8.369e-3	1. 880e-2	1.303e-2
	0.02	6.256e-1	6.733e-3	1.669e-2	2.631e-2	2.565e-2
	0.05	★★	1.702e-2	4.277e-2	3.682e-2	6.345e-2
0.6	0.01	1.244e-1	6.247e-3	3.395e-2	1.232e-2	1.363e-2
	0.02	5.707e-1	1.245e-2	3.522e-2	2.545e-2	2.845e-2
	0.05	★★	3.273e-2	5.614e-2	7.845e-2	6.227e-2
0.8	0.01	7.477e-2	8.407e-3	1.478e-2	1.501e-2	1.458e-2
	0.02	1.498e-1	1.718e-2	3.342e-2	2.931e-2	2.915e-2
	0.05	2.540e-1	4.192e-2	9.604e-2	8.648e-2	8.705e-2
1.0	0.01	8.008e-2	1.044e-2	2.763e-2	1.509e-2	1.636e-2
	0.02	1.072e-1	2.097e-2	7.136e-2	3.024e-2	3.421e-2
	0.05	1.599e-1	5.289e-2	2.083e-1	6.722e-2	7.504e-2

Table 4.38 - Modal loss factors for different material loss factors (η) and thickness to half wavelength numbers (torsional mode $n = 0$). ($H/R = 0.3$)

		Modal loss factor		
H/L	η	1	2	3
0.0	0.01	★	1.019e-2	2.045e-2
	0.02	★	2.084e-2	4.022e-2
	0.05	★	4.978e-2	9.826e-2
0.1	0.01	★★	1. 024e-2	2.038e-2
	0.02	★★	2.073e-2	4.037e-2
	0.05	★★	4.994e-2	7.059e-2
0.2	0.01	8.290e-3	1. 038e-2	2.042e-2
	0.02	1.141e-2	1.037e-2	2.019e-2
	0.05	★★	5.318e-2	8.694e-2
0.4	0.01	7.764e-3	1. 096e-2	1.961e-2
	0.02	9.278e-3	2.215e-2	4.112e-2
	0.05	★★	5.287e-2	7.432e-2
0.6	0.01	6.739e-3	1. 183e-2	2.109e-2
	0.02	2.117e-4	2.465e-2	4.222e-2
	0.05	5.034e-4	5.621e-2	6.938e-2
0.8	0.01	1. 809e-4	1.293e-2	2.135e-2
	0.02	3.740e-4	2.645e-2	4.277e-2
	0.05	4.241e-2	7.431e-2	6.724e-2
1.0	0.01	2.956e-4	1. 427e-2	2.290e-2
	0.02	5.599e-4	3.759e-2	4.401e-2
	0.05	4.831e-2	7.176e-2	7.694e-2

Table 4.39 presents the first natural frequency factors for axial bending, associated with different circumferential numbers, for two different simply supported thin elastic cylindrical shells. This table depicts a comparison of the presented solution and those extracted from Armenakas et al. (1969) and Markus, (1988), as well as, the reported results using the Flugge's thin shell theory (see Markus 1988). The non-dimensionalized frequency factor utilized by Markus (1988) is of a different form than the frequency factor introduced in this text and it is presented as:

$$\Omega' = R\omega\sqrt{\frac{\rho}{E}(1-\nu^2)} \tag{4.13}$$

To convert the frequency factor used by Markus (1988), Ω', to the one introduced in this text the following equation can be used.

$$\Omega = 0.538041909\Omega' \frac{H}{R} \tag{4.14}$$

For the results presented in Table 4.31 the frequency factor of Ω' is used. These results are for two cases of thin cylindrical shells with $H/R = 0.05$ and $H/R = 0.02$ in combination with $R/L = 0.05$ (long shell).

Markus (1988) inferred the following four conclusions from the comparison with those of Flugge's:

1. Shell theory always produces values of Ω' larger than those based on theory of elasticity.

2. The observed relative error never exceeded 10% and therefore differences in the values obtained in the two theories are negligible for very thin cylinders.

3. The approximate theories are satisfactory for thin cylinders from the engineering viewpoint.

4. Flugge's results fall within 0.01% of all other investigators using other versions of the shell theory.

As shown, the results reported by the present method and those of Markus (1988) are in relatively similar. In particular, the results presented by Armenakas et al. (1969) are consistently higher than those of Markus. It should be noted that results listed under Armenakas et al. for the long cylinder have been linearly interpolated from their publication. Obviously, one should expect an appreciable error by using such a technique to interpolate data, mainly for lower frequency factors. However, for higher modes the error is reduced. It should be noted that Armenakas et al. (1969) did not provide results for the extremely thin cylinder.

Table 4.39 - Comparison of Fundamental Frequency Parameters for a thin long cylinder $R/L = 0.05$.

		Resonant Frequency Factors, Ω			
H/R	n	Flugge	Armenakas et al.	Markus	Present
0.002	0	0.096191	No Data	0.092930	0.092930
	1	0.016352	No Data	0.016101	0.016101
	2	0.005529	No Data	0.005452	0.005446
	3	0.005060	No Data	0.005037	0.005046
	4	0.008538	No Data	0.008534	0.008531
0.05	0	0.096191	0.149198	0.092929	0.092929
	1	0.016349	0.023820	0.016106	0.016095
	2	0.392916	0.040889	0.039233	0.039235
	3	0.109782	0.109842	0.109477	0.109482
	4	0.210197	0.209091	0.209008	0.209007

4.7 Remarks

In general, the natural frequency factors presented in this chapter for thick circular cylinders are very similar with those given by Armenakas et al. (1969). Nevertheless, in some cases, Armenakas' results slightly differ from the presented computed results. These deviations occur mostly in the fifth digit after the decimal point for the computed frequency factors and are believed to be caused by the inaccuracy of the techniques used for the constructions of Bessel and modified Bessel functions. Furthermore, it should be noted that due to the nature of Bessel functions of the second kind, for very low frequency factors and very short wavelength ratios, they are not consistently accurate in these conditions.

4.8 Key Symbols

A	stress amplitude
E	Young's modulus
E'	real parts of E
F	direct or shear stresses
G, μ	shear modulus, modulus of rigidity
G'	real parts of G
H	thickness of the cylinder
L	axial half wavelength
n	wave number
$n = 0$	breathing mode
$n = 1$	rigid body mode
$n > 1$	lobar modes
r	radial direction
$r = a$	inner surface boundary
$r = b$	outer surface boundary
R	mean radius of the cylinder
t	time
u_r, u_θ, u_z	displacements in r, θ, z directions
$\mathbf{X}$	vector of constants, vector of unknowns
z	axial direction
η	loss-factor
θ	transverse direction
λ	Lame's elastic constant
μ, G	shear modulus, modulus of rigidity
ν	Poisson's ratio
ρ	density
σ	normal stress
σ_{rr}, $\sigma_{\theta\theta}$, σ_{zz}	direct stress in the r, θ, z directions
τ	shear stress
$\tau_{r\theta}$	shear stress along θ and perpendicular to r
τ_{rz}	shear stress along z and perpendicular to r
$\tau_{\theta z}$	shear stress along z and perpendicular to θ
ω	excitation frequency
ω_s	the lowest simple thickness shear frequency
Ω	frequency factor
$\zeta = \pi/L$	axial wave number

5

Vibration of Multi-Layer thick cylinders

Free vibration of a laminated multi-layer, thick infinitely long cylindrical structure is presented by adopting the analytical solution for a single-layer cylinder. The solution is achieved by determining the displacements and stresses for each interface and by complying with requirements at the interfaces. A propagator matrix relating the boundary displacements to boundary stresses is developed. Dimensionless natural frequencies and modal loss factors for different circumferential and axial wave numbers are determined. The validity of the proposed method is verified by comparing the results for one-, two-, and three-layer elastic cylinders with properties similar to those reported for an equivalent single layer.

5.1 Introduction

Laminated cylinders have widespread use in engineering. Pipe systems, aircraft, submarines, missiles, rockets, and power transmission shafts are typical applications for this type of cylindrical structure. When the frequency of an acting force on a cylinder coincides with the natural frequency of the structure, the phenomenon of resonance occurs. Resonance is associated with large displacements and stresses, and may cause failure in the cylinder. Therefore, determining the natural frequency is vital and has a direct impact on the design of composite cylindrical structures.

To control large displacements and stresses, damping treatment using visco-elastic material with a complex shear modulus are employed. In constrained layer damping treatment, a visco-elastic material is sandwiched between two elastic layers. The visco-elastic layer dissipates much of the energy within the visco-elastic cylinder in the form of heat during shear deformation, when subjected to harmonic vibration. In this configuration, the middle layer allows shear displacement, which converts dissipation of energy into heat, while the elastic layers give the cylindrical structure rigidity and stability.

Many researchers have investigated the vibration of cylinders using different analytical approaches, such as: shell theory, curved beam theory, and

H.R. Hamidzadeh, R.N. Jazar, *Vibrations of Thick Cylindrical Structures*
DOI 10.1007/978-0-387-75591-5_5,

theory of elasticity. Most reported studies of the vibrations in constrained layer damped cylinders are based on the assumptions for thin shell cylinders. The assumptions used in shell theory for cylindrical structures may not always be valid. In fact, a significant number of cylindrical structures are considered to be thick. In spite of this, few studies have been directed to the analysis of free vibrations of thick cylindrical structures.

5.2 Historical Background

In the foregoing section, a summary of notable research work on this topic is presented. Initial research on vibrations of sandwich cylindrical shells was conducted by White (1960) and Yu (1960), who considered visco-elastic properties for three-layered cylindrical structures. Jones and Salerno (1965) investigated the damping effect in the forced axisymmetrical vibration of a cylindrical sandwich shell. The sandwich cylinders considered by Yu (1960), Jones and Salerno (1965) were infinitely long or simply supported.

Pan (1968) studied axisymmetrical vibrations of a circular sandwich shell using a visco-elastic layer with a complex shear modulus. He derived and presented the partial differential equations for the motion. The solution of these differential equations yields the frequency equation and the corresponding modal loss-factor after satisfying the boundary conditions. Lu et al. (1973) introduced damping material in the middle layer and analyzed the mechanical impedance of damped three-layered sandwich rings by using the approximate method. They considered the effects of the operational temperature and frequency ranges on the visco-elastic material. The experimental data compared reasonably well with their theoretical predictions for lower modes. Hamidzadeh and Chandler (1991) analytically simulated one of the cases presented by Lu et al. (1973) using forced vibration analysis. Their results compared satisfactorily with those of Lu et al. (1973).

Hamidzadeh (1990) presented solutions for lobar vibration of three layered sandwiched cylinders of infinite extend. Hamidzadeh and Chandler (1991) also reported a solution to circumferential vibrations of three-layered, sandwiched, thick cylinders using the plane strain assumptions.

Rattanawangcharoen and Shah (1990) studied wave propagation in a laminated isotropic cylinder. The dispersion relation for a multi-layered elastic cylinder was evaluated by a propagator matrix and non-dimensional natural frequencies for a two-layered cylinder for a special circumferential wave number, $n = 1$, and a range of axial wave numbers were calculated.

Hawkes and Soldatos (1992) studied vibrations of multi-layered laminated cylinders. They achieved the solution using a method of successive approximations. In their approach, the exact governing equations for the cylinders were replaced by a set of approximate equations that were solved analytically. They compared their results with those of Armenakas et al. (1969).

Hamidzadeh and Sawaya (1995) presented results of their analysis of constrained layer damping of thick cylinders. They developed a propagator matrix relating displacements and stresses at the interface of a layer to its adjacent layer. They introduced their analytical solution for determining the natural frequencies and modal loss-factors for a number modes of vibrations for the three-layer, thick, sandwiched cylinders of infinite length. Hamidzadeh and Jiang (1995) used the same analytical method and presented modal loss-factors for three-layered sandwiched thick cylinders in a number of case studies. Hamidzadeh (2009) adopted the same analytical method and considered the effect of visco-elastic core thickness on modal loss-factor of a thick three-layer cylinder.

5.3 Scope of the Chapter

The governing equation for thick, axially symmetric, as well as non-symmetric isotropic visco-elastic multi-layered cylinders of infinite extent will be presented. The three-dimensional analytical solution developed in chapter three using wave propagation, which applies equally to thick or thin cylinders, will be applied. The method presented here does not use any approximation. It relates the modal displacements and stresses of the first layer to the last layer in the cylinder using a propagator matrix. The advantage of this method is that it can deal with any number of layers. The method is discussed in detail in this chapter.

As reported by Hamidzadeh and Sawaya (1995), the solution is achieved by determining the displacements and stresses for each interface and by complying with the compatibility requirements at the boundaries. A propagator matrix relating boundary displacements and stresses between the inner and outer layers is developed. This method can consider any number of elastic and visco-elastic layers for a cylinder.

Non-dimensional natural frequencies and modal loss-factors for different circumferential and axial wave numbers are calculated for both axial and nonaxisymmetic modes. Case studies and the required computation were conducted to determine the natural frequencies for single-layer, two-layer, and three-layer cylinders. Results of these case studies are compared with those reported by Armenakas et al. (1969) for elastic cylinders with similar geometries. This was done to verify the accuracy of the analytical method relating the inner layer to the outer one (Hamidzadeh and Sawaya, 1995). The analytical solutions presented in chapter 3 are used to develop the analytical solutions for determination of modal characteristics for fully covered, three-layer, and multi-layer cylindrical structures.

5.4 Modal Displacements and Stresses

As was presented in chapter three, using equations (3.108) and (3.84), modal displacements and stresses can be written in terms of Bessel functions in the following equation:

$$\begin{Bmatrix} u_r/h_1 \\ u_\theta/h_2 \\ u_z/h_3 \\ \sigma_{rr}/h_1 \\ \tau_{r\theta}/h_2 \\ \tau_{rz}/h_3 \end{Bmatrix} = \begin{bmatrix} D_{1,1} & D_{1,2} & D_{1,3} & D_{1,4} & D_{1,5} & D_{1,6} \\ D_{2,1} & D_{2,2} & D_{2,3} & D_{2,4} & D_{2,5} & D_{2,6} \\ D_{3,1} & D_{3,2} & D_{3,3} & D_{3,4} & D_{3,5} & D_{3,6} \\ T_{1,1} & T_{1,2} & T_{1,3} & T_{1,4} & T_{1,5} & T_{1,6} \\ T_{2,1} & T_{2,2} & T_{2,3} & T_{2,4} & T_{2,5} & T_{2,6} \\ T_{3,1} & T_{3,2} & T_{3,3} & T_{3,4} & T_{3,5} & T_{3,6} \end{bmatrix} \begin{Bmatrix} A_1 \\ B_1 \\ A_2 \\ B_2 \\ A_3 \\ B_3 \end{Bmatrix} \tag{5.1}$$

where elements of the matrix are in terms of the Bessel functions and were presented in chapter 3. It should be noted that:

u_r is the displacement in the r direction,
u_θ is the displacement in the θ direction,
u_z is the displacement in the z direction,
σ_{rr} is the normal radial stress in the r direction,
$\tau_{r\theta}$ is the shear stress in the θ direction and normal to r,
τ_{rz} is the shear stress in the z direction and normal to r, and
$\mathbf{X}$ is the constant vector given previously in equation (3.83)

Then equation (5.1) can be presented as:

$$\{d\}_n = [D]_n \{X\}_n \tag{5.2}$$

where, the vector $\{d\}_n$ is:

$$\{d\}_n = \left\{ \frac{u_r}{h_1} \quad \frac{u_\theta}{h_2} \quad \frac{u_z}{h_3} \quad \frac{\sigma_{rr}}{h_1} \quad \frac{\sigma_{r\theta}}{h_2} \quad \frac{\sigma_{rz}}{h_3} \right\}_n^T \tag{5.3}$$

and $[D]_n$ previously appeared in equation (3.107), and n is the circumferential wave number.

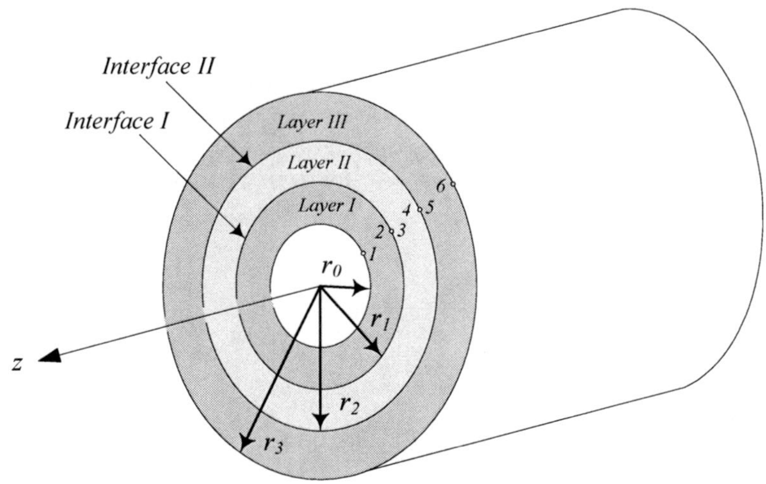

FIGURE 5.1. Cross Section of a Three Layered Cylinder.

5.5 Propagator Matrix for a Three Layer Cylinder

The layered cylinder is assumed to be formed by three different visco-elastic layers bounded by inner and outer interfaces (see Figure 5.1). The third layer has a thickness of $H_3 = r_3 - r_2$ and is bounded by inner and outer interfaces located at radii of r_2 and r_3, respectively. Each of the visco-elastic cylinders is characterized by a complex compressional wave velocity v_1, complex shear wave velocity v_2, density ρ, Poisson's ratio ν, and a material loss-factor η.

The mathematical formulation is used to develop a relation between displacements and stresses at the boundaries of each layer and consequently obtains a propagator matrix relating displacements and stresses of the inner boundaries to the outer ones. To build the propagator matrix for a three layered cylinder, the relations for stresses and displacements at any location in the cylindrical structure, defined by equation (5.1) is used.

In Figure (5.1), points (1) and (2), (3) and (4), and (5) and (6), lay on layer I, layer II and layer III, respectively. All points laying on the same layer comprise the same unknown vector $\{X\}_n$. Points (2) and (3), and (4) and (5), lay on interface I and II, respectively. Therefore, points (2) and (3), and (4) and (5), have the same displacements and stresses.

As presented in equation (5.2), the displacements and stresses vector for any point on the cylinder can be given as follows:

$$\{d\}_{np} = [D]_{np} \{X\}_{np} \tag{5.4}$$

where, n is the circumferential wave number and p is the point number.

Using the general equation of (5.1), the following equations can be written for each point in the medium. Writing the displacements and stresses vector for the six points presented in Figure 5.1 and dropping subscript n for simplification one can write the following equations for points (1) through (6).

point (1):

$$\{d\}_1 = [D]_1 \{X\}_1 \tag{5.5}$$

point (2):

$$\{d\}_2 = [D]_2 \{X\}_2 \tag{5.6}$$

point (3):

$$\{d\}_3 = [D]_3 \{X\}_3 \tag{5.7}$$

point (4):

$$\{d\}_4 = [D]_4 \{X\}_4 \tag{5.8}$$

point (5):

$$\{d\}_5 = [D]_5 \{X\}_5 \tag{5.9}$$

point (6):

$$\{d\}_6 = [D]_6 \{X\}_6 \tag{5.10}$$

Points (1) and (2) are on the same layer, therefore:

$$\{X\}_1 = \{X\}_2 \tag{5.11}$$

Therefore, using these equalities and equations (5.5) and (5.6) yield:

$$\{d\}_2 = [D]_2 [D]_1^{-1} \{d\}_1 \tag{5.12}$$

Points (3) and (4), and (5) and (6), are on layers two and three respectively, this implies that points (3) and (4), and (5) and (6), have the same unknown constants, therefore:

$$\begin{aligned} \{X\}_3 &= \{X\}_4 && (5.13) \\ \{X\}_5 &= \{X\}_6 && (5.14) \end{aligned}$$

Using the same procedure used for the first layer which resulted the equation (5.12), one can write the following two equations for the second and the third layers.

$$\begin{aligned} \{d\}_4 &= [D]_4 [D]_3^{-1} \{d\}_3 && (5.15) \\ \{d\}_6 &= [D]_6 [D]_5^{-1} \{d\}_5 && (5.16) \end{aligned}$$

Points (2) and (3) lay on different layers but they have the same displacement and stresses. Similarly, points (4) and (5) are on different layers but have the same displacement and stresses. This implies that:

$$\{d\}_2 = \{d\}_3 \qquad \{d\}_4 = \{d\}_5 \tag{5.17}$$

Substituting $\{d\}_3$ for $\{d\}_2$ in (5.12), it yields:

$$\{d\}_3 = [D]_2\,[D]_1^{-1}\,\{d\}_1 \tag{5.18}$$

Similarly, one can write:

$$\{d\}_5 = [D]_4\,[D]_3^{-1}\,\{d\}_3 \tag{5.19}$$

Introducing $[P_I]$, $[P_{II}]$, and $[P_{III}]$ according to the following expressions:

$$[P_I] = [D]_2\,[D]_1^{-1} \tag{5.20}$$

$$[P_{II}] = [D]_4\,[D]_3^{-1} \tag{5.21}$$

$$[P_{III}] = [D]_6\,[D]_5^{-1} \tag{5.22}$$

Combining equations (5.16), (5.18), and then (5.19), the relationship between point (1) and point (6) is established and is equal to:

$$\{d\}_6 = [P_{III}]\,[P_{II}]\,[P_I]\,\{d\}_1 \tag{5.23}$$

Introducing $[P]$ as:

$$[P] = [P_{III}]\,[P_{II}]\,[P_I] \tag{5.24}$$

Therefore,

$$\{d\}_6 = [P]\,\{d\}_1 \tag{5.25}$$

For free vibrations, the stresses at the inner and outer radii are equal to zero. Therefore equation (5.25) can be written as:

$$\left\{\begin{array}{c} \frac{u_r}{h_1} \\ \frac{u_\theta}{h_2} \\ \frac{u_z}{h_3} \\ 0 \\ 0 \\ 0 \end{array}\right\}_6 = \begin{bmatrix} P_{1,1} & P_{1,2} & P_{1,3} & P_{1,4} & P_{1,5} & P_{1,6} \\ P_{2,1} & P_{2,2} & P_{2,3} & P_{2,4} & P_{2,5} & P_{2,6} \\ P_{3,1} & P_{3,2} & P_{3,3} & P_{3,4} & P_{3,5} & P_{3,6} \\ P_{4,1} & P_{4,2} & P_{4,3} & P_{4,4} & P_{4,5} & P_{4,6} \\ P_{5,1} & P_{5,2} & P_{5,3} & P_{5,4} & P_{5,5} & P_{5,6} \\ P_{6,1} & P_{6,2} & P_{6,3} & P_{6,4} & P_{6,5} & P_{6,6} \end{bmatrix} \left\{\begin{array}{c} \frac{u_r}{h_1} \\ \frac{u_\theta}{h_2} \\ \frac{u_z}{h_3} \\ 0 \\ 0 \\ 0 \end{array}\right\}_1 \tag{5.26}$$

The above equation yields to:

$$\begin{bmatrix} P_{4,1} & P_{4,2} & P_{4,3} \\ P_{5,1} & P_{5,2} & P_{5,3} \\ P_{6,1} & P_{6,2} & P_{6,3} \end{bmatrix} \left\{\begin{array}{c} \frac{u_r}{h_1} \\ \frac{u_\theta}{h_2} \\ \frac{u_z}{h_3} \end{array}\right\}_1 = \left\{\begin{array}{c} 0 \\ 0 \\ 0 \end{array}\right\} \tag{5.27}$$

For the nontrivial solutions of the above equation, the determinant of the 3×3 matrix must be equal to zero and the roots of the determinant are the natural frequency of the composite cylinder. In other words, the determinant of the coefficient matrix in equation (5.27) provides the frequency equation for the system. It should be indicated that elements of the above matrix are all related to the circumferential and axial wave numbers, the material loss-factor, and the geometry of each layer. Dimensionless natural frequencies and modal loss-factors for different modes can be computed using Muller's iterative method as described by Conte (1965).

5.6 Propagator Matrix for a Multi-Layered Thick Cylinder

In summary, a multi-layered composite cylinder for a set of a specific wave number can be formulated in the manner described in the previous section. Considering Figure 5.2, which presents a multi-layered composite cylinder for the ith and $(i + l)th$ interface, we can write:

$$\{d\}_i = [D]_i \{X\}_i \tag{5.28}$$

$$\{d\}_{i+1} = [D]_{i+1} \{X\}_{i+1} \tag{5.29}$$

Since for the ith layer $\{X\}_i$ is the same as $\{X\}_{i+1}$, one can write the following equation:

$$\{d\}_{i+1} = [D]_{i+1} [D]_i^{-1} \{d\}_i \tag{5.30}$$

or

$$\{d\}_{i+1} = [P]_{i+1} \{d\}_i \tag{5.31}$$

where,

$$[P]_{i+1} = [D]_{i+1} [D]_i^{-1} . \tag{5.32}$$

Writing equation (5.31) for all layers and eliminating vector $\{d\}$ for all intermediate interfaces results in a relation between displacements and stresses vectors at the inner and outer boundaries of the sandwiched cylinder.

$$\{d\}_{i+1} = [P]_{i+1} [P]_i [P]_{i-1} \cdots [P]_1 \{d\}_1 \tag{5.33}$$

It can be rewrite as

$$\{d\}_{i+1} = [P] \{d\}_1 \tag{5.34}$$

where,

$$[P] = [P]_{i+1} [P]_i [P]_{i-1} \cdots [P]_1 . \tag{5.35}$$

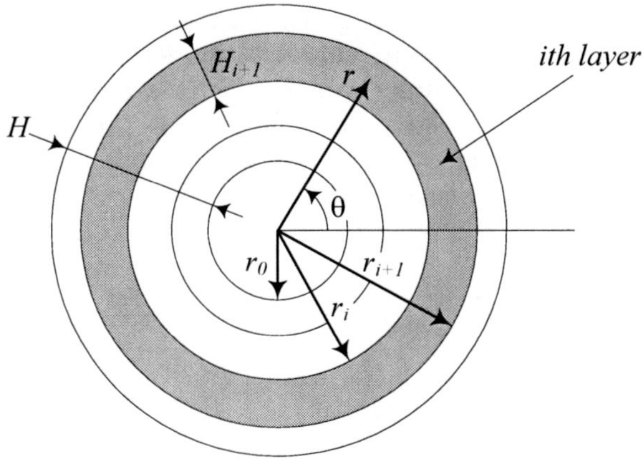

FIGURE 5.2. Cross Section of a Multi-Layered Circular Cylinder.

The 6×6 matrix $[P]$ is the propagator matrix. For free vibration, as was presented for the three layered cylinder equation (5.34) reduces to:

$$\begin{bmatrix} P_{4,1} & P_{4,2} & P_{4,3} \\ P_{5,1} & P_{5,2} & P_{5,3} \\ P_{6,1} & P_{6,2} & P_{6,3} \end{bmatrix} \begin{Bmatrix} \dfrac{u_r}{h_1} \\ \dfrac{u_\theta}{h_2} \\ \dfrac{u_z}{h_3} \end{Bmatrix}_1 = \begin{Bmatrix} 0 \\ 0 \\ 0 \end{Bmatrix} \tag{5.36}$$

5.7 Natural Frequencies and Modal Loss-Factors

The nontrivial solutions of equation (5.27) for three-layer cylinders or (5.36) for multi-layer cylinders requires that:

$$\begin{vmatrix} P_{4,1} & P_{4,2} & P_{4,3} \\ P_{5,1} & P_{5,2} & P_{5,3} \\ P_{6,1} & P_{6,2} & P_{6,3} \end{vmatrix} = 0 \tag{5.37}$$

This equation provides the natural frequencies of the composite sandwich cylinders with infinte extend. Since elements of the matrix are in terms of Bessel functions, the equation (5.39) has an infinite number of solutions for any circumferential wave number. In this text, an attempt is made to determine only the first six roots of equation (5.39). As mentioned in chapter four, Muller's method which is an extension of the secant method is used. This method may be used to find any prescribed number of roots, real or complex, of an arbitrary function. Muller's method computes both real and complex roots without requiring any derivative of the function. A detailed description of Muller's method is presented by Conte (1965), Beckett and Hurt (1967), Conte and Boor (1980). The roots of equation (5.37) are natural frequencies of the layered cylinders and are in complex form for visco-elastic materials:

$$\omega = \Re\omega + i\Im\omega \tag{5.38}$$

where $\Re\omega$ =real part of ω and $\Im\omega$ =imaginary part of ω. The above equation can also be presented as:

$$\omega = \Re\omega\,(1 + i\alpha) \tag{5.39}$$

where α is the ratio of the imaginary over the real part of the root.

For visco-elastic materials, the modal loss-factor η can be defined as:

$$\eta = \frac{\Im\omega^2}{\Re\omega^2} \tag{5.40}$$

In this text, the frequency factor, Ω, as defined by Armenakas et al. (1969), is implemented, which is:

$$\Omega = \frac{\Re\omega}{\omega_s} \tag{5.41}$$

where,

$$\omega_s = \frac{\pi v_2}{H} \tag{5.42}$$

and v_2 is the velocity of the distortional wave for the inner layer

$$v_2 = \sqrt{\frac{\mu}{\rho}} \tag{5.43}$$

and H is the total thickness of the layered cylinder.

5.8 Sample Calculations for One Layered Cylinder

Consider a single layer steel cylinder with

thickness of $H = 0.35294\,\text{in}$,
density of $\rho = 0.000725\,\text{lb}\,\text{s}^2/\,\text{in}^4$, and
shear modulus of $\mu = 1.1538 \times 10^7\,\text{lb}/\,\text{in}^2$.

Figure 5.3 depicts the geometrical properties of the considered single layer cylinder, which has the following specifications:

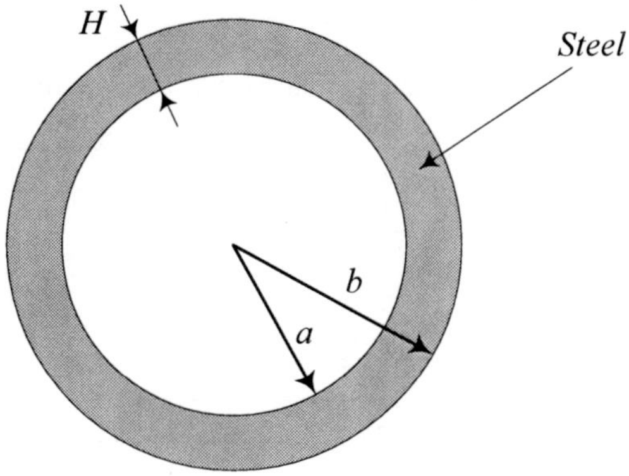

FIGURE 5.3. The geometrical properties of a cylinder.

inner radius: $a = 1\,\text{in.}$
outer radius: $b = 1.35294\,\text{in.}$
mean radius: $R = (a+b)/2 = (1+1.35294)/2 = 1.17650\,\text{in.}$
thickness of the cylinder H: $H = b - a = 1.35294 - 1 = 0.35294\,\text{in.}$
thickness to radius ratio of H/R: $H/R = 0.35294/1.17650 = 0.3$.

Assuming the thickness to length $H/L = 0.1$. The axial wave length ζ was defined in section 4.4 by equation (4.8) and in this case can be written as: is analyzed.

$$\zeta H = \frac{\pi H}{L} \tag{5.44}$$

$$\zeta = \frac{0.1\pi}{0.35294} = 0.890\,12 \tag{5.45}$$

Therefore, the half wave length is: $L = \pi/\zeta \approx 3.53\,\text{in.}$

Utilizing the above data and satisfying the boundary condition yields the 3×3 matrix of equation (5.36). Using Muller's method, the roots of the determinant of the matrix were found. To calculate the frequency ω, the non-dimensionalized frequency must be multiplied by the following factor as shown.

$$\Re\omega = \Omega\omega_s = \Omega\frac{\sqrt{\dfrac{1.1538 \times 10^7}{0.725 \times 10^{-3}}}}{0.35294}\pi = 1.122\,9 \times 10^6\Omega \tag{5.46}$$

and the natural frequency in Hz can be written as:

$$f = \frac{1.122\,9 \times 10^6}{2\pi}\Omega = 1.7872 \times 10^5\Omega \tag{5.47}$$

The first six natural frequencies for this specific case were computed using equations (5.39) and (5.42) and are presented in Table 5.1.

Table 5.1 - Natural frequencies for a steel cylinder with $a = 1$ in, $b = 1.35294$ in, and $H/L = 0.1$, for $n = 0$.

	Nondimensionalized Frequencies	Frequencies, Hz
Mode	Ω	f
First	0.13828	24713
Second	0.18945	33859
Third	1.02108	182490
Fourth	1.85679	331850
Fifth	2.02624	362130
Sixth	3.00083	536310

As seen in this Table 5.1 the sixth mode is on the order of 0.5 MHz, which is a very high frequency for cylindrical structures. Only specific applications can reach high frequencies as high as 5 MHz. Therefore, the sixth mode will meet most mechanical applications.

5.9 Modal Identification

As defined in a previous chapter (Chapter 4), the modes of vibration for a cylinder are as follows:

For $n = 0$, the lowest mode is the torsional, followed by the lowest extensional mode and the breathing mode. Then comes a circumferential shear (torsional) mode, an axial shear mode, and a thickness-stretch mode; all three of which are characterized by one cylindrical nodal surface near the middle of the thickness.

For $n \geq 1$, the lowest mode is the flexural (bending) mode, associated with large radial displacements. The second (higher) mode is an axial shear mode, associated with large z displacements. The third is a breathing mode, with large transverse displacements, and the fourth and fifth are thickness-shear modes associated with motions in the z and θ directions, respectively. The sixth mode is a thickness-stretch mode with predominantly radial displacement.

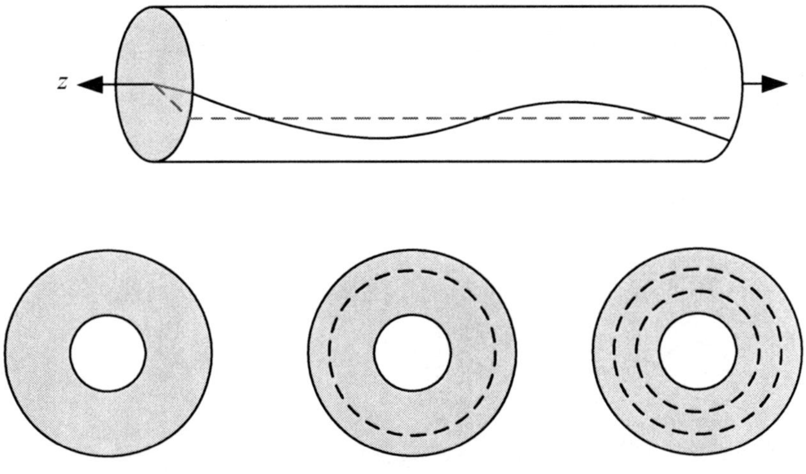

FIGURE 5.4. Torsional Mode and the First Three Related Thickness Modes, $n = 0$.

5.9.1 Axially Symmetric and Asymmetric Vibrations

Axially symmetric vibration occurs only in breathing mode with radial displacement (Markus 1988). Longitudinal modes occur with axial displacement, and torsional modes with transverse displacement around the circumference of the cross section. Figures 5.4 through 5.6 depict the three axisymmetric modes associated with the corresponding first three thickness modes of vibrations.

5.9.2 Thickness Modes

Vibration of cylinders is mathematically composed of an infinite number of modes. For each of the torsional, axial, breathing, bending, and lobar modes of vibration, there are an infinite number of solutions. This is due to the nature of differential equations and is manifested by a finite number of enclosed nodal lines on the cross section of the cylindrical structure. Therefore, for free vibrations of cylinders, there are an infinite number of solutions. In this chapter, only the first six natural frequencies associated with thickness modes are computed. Figures 5.4 through 5.6 show the first three thickness modes for different axially symmetric vibrations.

It should be noted that each torsional mode of vibration is associated with a periodical motion of points on the cross section in the direction of θ only, where u_r and u_z are zero. For these modes, the enclosed nodal lines are in the form of circles within the thickness of the cylinder and have the

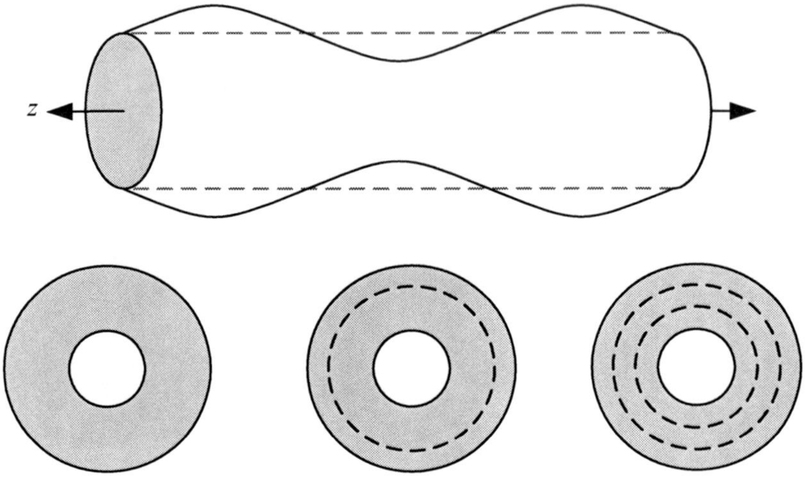

FIGURE 5.5. Radial Mode and the First Three Thickness Modes, $n = 0$.

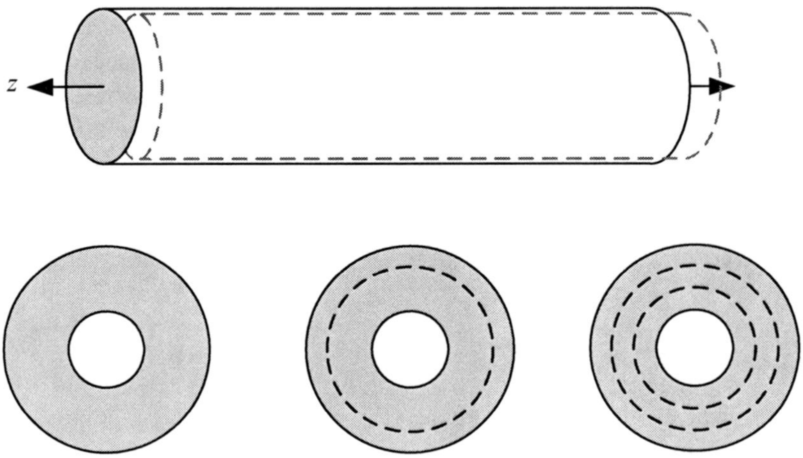

FIGURE 5.6. Longitudinal Mode and the First Three Thickness Modes, $n = 0$.

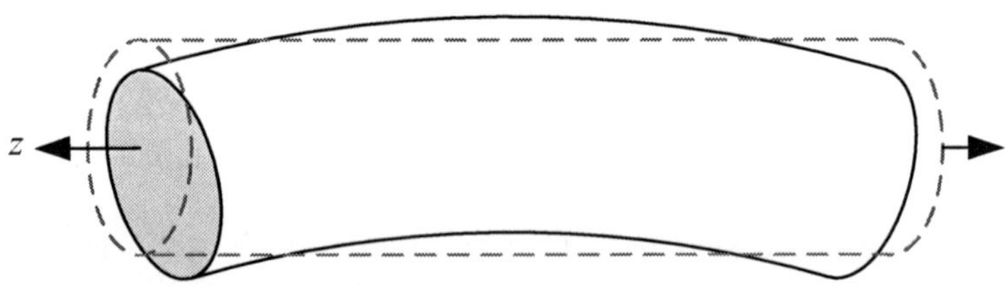

FIGURE 5.7. Bending mode for $n = 1$.

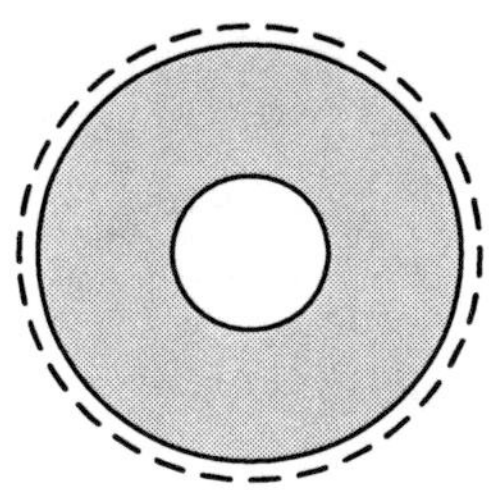

FIGURE 5.8. Breathing Mode, $n = 0$.

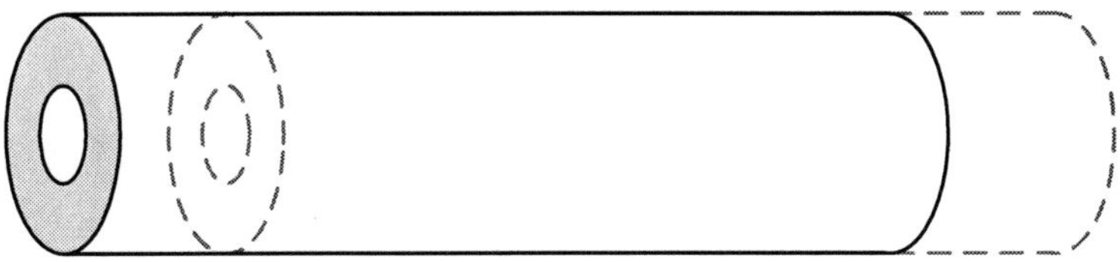

FIGURE 5.9. Rigid Body Mode, $n = 1$.

same center point as the cross-section. For each of the pure longitudinal modes, vibration is associated with the periodical motion of points within the medium along the direction of the z axis only in the absence of u_r and u_θ. For these modes, the enclosed nodal lines are also formed with the shape of the circles as described for the pure torsional modes. Similarly, for the breathing modes, the only non-zero motion is in the radial direction with circular nodal lines.

The flexural modes are associated with large radial and transverse displacement amplitudes varying along the length and the circumference of the cylindrical structure. These modes occur for $n > 1$. For the pure bending vibration mode along the axis, the circumferential number is equal to 1 ($n = 1$). This mode is depicted in Figure 5.7.

The modes of vibration for the multi-layer cylinder are commonly defined as Lobar modes for $n > 1$, bending and axial shear modes for $n = 1$, and the breathing, longitudinal, and torsional modes for $n = 0$. In this chapter, individual modes for $n = 0, 1, 2, 3$, and 4 are examined. These modes are the dominant modes of vibrations. The dominance of a particular mode depends upon the type of excitation. Higher modes are excited at very high frequencies and are usually not significant. These five dominant modes are shown in Figures 5.7 to 5.12.

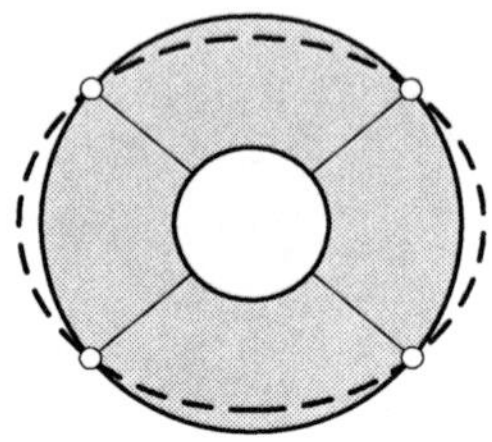

FIGURE 5.10. First Lobar Mode, $n = 2$.

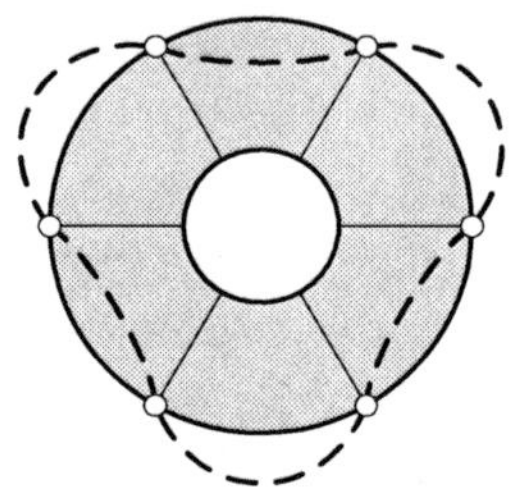

FIGURE 5.11. Second Lobar Mode, $n = 3$.

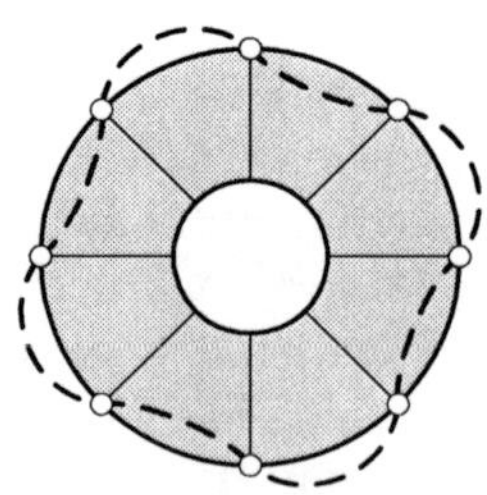

FIGURE 5.12. Third Lobar Mode, $n = 4$.

5.10 Case Studies

To verify the validity of the developed procedure, several case studies are presented in the following sections.

5.10.1 Elastic Cylinder

Dimensionless natural frequencies are computed for the following parameters and are compared with those reported by Armenakas et al. (1969)

circumferential wave numbers ($n = 0, 1, 2, 3,$ and 4),
$H/L = 0, 0.1, 0.5, 1,$
$H/R = 0.3,$ and
Poisson's ratio $= 0.3$

The parameters H, R, and L are thickness, mean radius, and half wave length of the cylinder, respectively. These results are tabulated in Table 5.2 and comparison indicates excellent agreement. Dimensionless natural frequencies are defined as $\Omega_j = \omega_j H/(\pi v_2)$, where ωj is the jth natural frequency and v_2 is the velocity of the shear wave for the cylinder.

Table 5.2. Comparison of dimensionless natural frequencies for a single elastic layer, $H/R = 0.3$, Ref [Armenakas et al.]

		Dimensionless Natural Frequencies						
n	H/L		Ω_1	Ω_2	Ω_3	Ω_4	Ω_5	Ω_6
	0.0	Present	0.00000	0.16305	1.00344	1.87870	2.00174	3.00116
		Ref.	0.00000	0.16306	1.00344	1.87871	2.00174	3.00118
	0.1	Present	0.13828	0.18945	1.02108	1.85679	2.02643	3.00083
0		Ref.	0.13828	0.18946	1.02055	1.85681	2.03137	3.00076
	0.5	Present	0.31801	0.82790	1.33439	1.74402	2.33664	3.00068
		Ref.	0.31790	0.82791	1.33441	1.74402	2.33665	3.00068
	1.0	Present	0.78333	1.42000	1.91154	2.00292	2.33665	3.07879
		Ref.	0.78471	1.42002	1.91156	2.00294	2.82864	3.07880
	0.0	Present	0.00000	0.09583	0.22521	1.00819	1.03402	1.85944
		Ref.	0.00000	0.09584	0.22524	1.00819	1.03404	1.85845
	0.1	Present	0.06158	0.16483	0.26024	1.01826	1.04543	1.84228
1		Ref.	0.06158	0.16484	0.26027	1.01826	1.04544	1.84229
	0.5	Present	0.32108	0.51064	0.84136	1.13768	1.34520	1.74345
		Ref.	0.32101	0.51065	0.84138	1.13767	1.34521	1.74344
	1.0	Present	0.73322	1.00433	1.43616	1.43616	1.91681	2.00869
		Ref.	0.78856	1.00483	1.41665	1.43618	1.91684	2.00872
	0.0	Present	0.03596	0.19158	0.35146	1.02229	1.08211	1.82135
		Ref.	0.03596	0.19161	0.35149	1.02230	1.08213	1.82136
	0.1	Present	0.05456	0.22783	0.38247	1.02966	1.09493	1.81191
2		Ref.	0.05456	0.22784	0.38247	1.02966	1.09492	1.81191
	0.5	Present	0.33333	0.54008	0.87974	1.15062	1.37682	1.74257
		Ref.	0.33332	0.54009	0.87975	1.15063	1.37683	1.74257
	1.0	Present	0.83333	1.01913	1.41767	1.45418	1.93217	2.02637
		Ref.	0.80015	1.01911	1.41769	1.45421	1.93221	2.02639

Table 5.2. Continued

			Dimensionless Natural Frequencies					
n	H/L		Ω_1	Ω_2	Ω_3	Ω_4	Ω_5	Ω_6
3	0.0	Present	0.09554	0.28718	0.49297	1.04542	1.15516	1.78653
		Ref.	0.09556	0.28721	0.49302	1.04543	1.15517	1.78652
	0.1	Present	0.10639	0.30949	0.51769	1.05171	1.16737	1.78117
		Ref.	0.10638	0.30951	0.51772	1.00517	1.16737	1.78116
	0.5	Present	0.35967	0.58365	0.93778	1.17221	1.42710	1.74382
		Ref.	0.35967	0.58365	0.93779	1.17220	1.42709	1.74382
	1.0	Present	0.72932	1.04213	1.42532	1.47738	1.95646	2.05673
		Ref.	0.81956	1.04226	1.42537	1.47741	1.95649	2.05677
4	0.0	Present	0.16994	0.38254	0.63690	1.07704	1.24634	1.75990
		Ref.	0.16997	0.38258	0.63697	1.07706	1.24635	1.75991
	0.1	Present	0.17916	0.39824	0.65666	1.08278	1.25734	1.75718
		Ref.	0.17917	0.39825	0.65666	1.08277	1.25739	1.75717
	0.5	Present	0.40148	0.63760	1.00884	1.20232	1.49258	1.75090
		Ref.	0.40149	0.63760	1.00895	1.20233	1.49260	1.75099
	1.0	Present	0.88660	1.07349	1.43943	1.50490	1.98824	2.10042
		Ref.	0.84673	1.07349	1.43945	1.50493	1.98827	2.10048

5.10.2 Two-Layer Elastic Cylinder

A two-layer elastic cylinder with the same mechanical properties is considered. Dimensionless natural frequencies for different circumferential wave numbers and two values of H/L are compared with those for an equivalent single-layer elastic cylinder. Results for two different geometries are given in Tables 5.3 and 5.4. Table 5.3 presents the results for $H/R = 0.3$ and $H/L = 0.1$ for a laminated cylinder with radii of $r_0 = 1.0\,\text{in}$, $r_1 = 1.15\,\text{in}$, and $r_2 = 1.3529\,\text{in}$. This is equivalent to a single layer of $H/R = 0.3$. Table 5.4 compares the dimensionless natural frequencies for two elastic laminated cylinders for:

$H/R = 0.5$,
$H/L = 0.5$,
$r_0 = 1.0\,\text{in}$, $r_1 = 1.1666\,\text{in}$, and $r_2 = 1.1666\,\text{in}$.

As indicated in these tables, computed results are very close to the results presented by Armenakas et al. (1969) for an equivalent single cylinder. Dimensionless natural frequencies for the multilayered cylinder are defined as $\Omega_j = \omega_j H/(\pi v_{21})$, where ω_j is the jth natural frequency and v_{21} is the velocity of the shear wave for the first layer.

Table 5.3. Comparison of dimensionless natural frequencies for two elastic layers, $H/R = 0.3$, $H/L = 0.1$, $r_0 = 1.0$ in, $r_1 = 1.15$ in, and $r_2 = 1.3529$ in, Ref [Armenakas et al.]

	Dimensionless Natural Frequencies						
n		Ω_1	Ω_2	Ω_3	Ω_4	Ω_5	Ω_6
0	Present	0.13828	0.18946	0.39825	1.85681	2.03136	3.00075
	Ref.	0.13829	0.18946	0.39825	1.85681	2.03136	3.00076
1	Present	0.06158	0.16484	0.26027	1.01826	1.04544	1.84228
	Ref.	0.06158	0.16484	0.26027	1.01826	1.04544	1.84229
2	Present	0.05456	0.22784	0.38247	1.02966	1.09492	1.81191
	Ref.	0.05456	0.22784	0.38247	1.02966	1.09492	1.81191
3	Present	0.10638	0.30951	0.51772	1.05170	1.16736	1.78117
	Ref.	0.10638	0.30951	0.51772	1.05170	1.16737	1.78116
4	Present	0.17917	0.39825	0.65665	1.08277	1.25739	1.75717
	Ref.	0.17917	0.39825	0.65665	1.08277	1.25739	1.75717

Table 5.4. Comparison of dimensionless natural frequencies for two elastic layers, $H/R = 0.5$, $H/L = 0.5$, $r_0 = 1.0$ in, $r_1 = 1.1666$ in, and $r_2 = 1.6666$ in, Ref [Armenakas et al.]

	Dimensionless Natural Frequencies						
n		Ω_1	Ω_2	Ω_3	Ω_4	Ω_5	Ω_6
0	Present	0.36913	0.83335	1.34362	1.75354	2.34374	3.00328
	Ref.	0.36919	0.83334	1.34362	1.75354	2.34374	3.00328
1	Present	0.35885	0.53632	0.86722	1.17272	1.37979	1.75152
	Ref.	0.35884	0.53632	0.86723	1.17273	1.37980	1.75153
2	Present	0.37309	0.61399	0.95428	1.21002	1.47866	75287
	Ref.	0.37308	0.61399	0.95428	1.21002	1.47867	75287
3	Present	0.44122	0.71206	1.06769	1.27295	1.61076	1.78279
	Ref.	0.44121	0.71206	1.06769	1.27295	1.61076	1.78280
4	Present	0.55006	0.82384	1.18781	1.35816	1.71423	1.89819
	Ref.	0.55006	0.82384	1.18781	1.35817	1.71424	1.89819

5.10.3 Three-layer Elastic Cylinder

Dimensionless natural frequencies for two different three-layer cylinders with the same elastic properties are compared with those of equivalent single cylinders, and are provided in Tables 5.5 and 5.6. Table 5.5 presents the results for:

$H/R = 0.3$,
$H/L = 0.1$,
$r_0 = 1.0\,\text{in}$, $r_1 = 1.15\,\text{in}$, $r_2 = 1.2\,\text{in}$, and $r_3 = 1.3529\,\text{in}$.

This is equivalent to a single layer of $H/R = 0.3$. Table 5.5 compares the dimensionless natural frequencies for three-layer cylinder with:

$H/R = 0.5$,
$H/L = 0.5$,
$r_0 = 1.0\,\text{in}$, $r_1 = 1.1666\,\text{in}$, $r_2 = 1.3333\,\text{in}$, and $r_3 = 1.6666\,\text{in}$.

The two tables show that the computed results are in excellent agreement with those reported by Armenakas et at. (1969) for an equivalent single cylinder.

Table 5.5. Comparison of dimensionless natural frequencies for three elastic layers, $H/R = 0.3$, $H/L = 0.1$, $r_0 = 1.0\,\text{in}$, $r_1 = 1.15\,\text{in}$, $r_2 = 1.2\,\text{in}$, and $r_3 = 1.35294\,\text{in}$, Ref [Armenakas et al.]

		Dimensionless Natural Frequencies					
n		Ω_1	Ω_2	Ω_3	Ω_4	Ω_5	Ω_6
0	Present	0.13828	0.18946	1.02055	1.85681	2.03136	3.00075
	Ref.	0.13829	0.18946	1.02055	1.85681	2.03137	3.00076
1	Present	0.06158	0.16484	0.26027	1.01826	1.04544	1.84228
	Ref.	0.06158	0.16484	0.26027	1.01826	1.04544	1.84229
2	Present	0.05456	0.22784	0.38247	1.02966	1.09492	1.81191
	Ref.	0.05456	0.22784	0.38247	1.02966	1.09492	1.81191
3	Present	0.10638	0.30951	0.51772	1.05170	1.16736	1.78117
	Ref.	0.10638	0.30951	0.51772	1.05170	1.16737	1.78117
4	Present	0.17917	0.39825	0.65665	1.08277	1.25739	1.75717
	Ref.	0.17917	0.39825	0.65666	1.08277	1.25739	1.75717

Table 5.6. Comparison of dimensionless natural frequencies for three elastic layers, $H/R = 0.5$, $H/L = 0.5$, $r_0 = 1.0\,\text{in}$, $r_1 = 1.1666\,\text{in}$, $r_2 = 1.3333\,\text{in}$, and $r_3 = 1.6666\,\text{in}$, Ref [Armenakas et al.]

	Dimensionless Natural Frequencies						
n		Ω_1	Ω_2	Ω_3	Ω_4	Ω_5	Ω_6
0	Present	0.36918	0.83335	1.34362	1.75354	2.34374	3.00328
	Ref.	0.36919	0.83334	1.34362	1.75354	2.34374	3.00328
1	Present	0.35883	0.53632	0.86723	1.17272	1.37979	1.75152
	Ref.	0.35884	0.53632	0.86723	1.17273	1.37980	1.75153
2	Present	0.37308	0.61399	0.95428	1.21002	1.47866	1.75287
	Ref.	0.37308	0.61400	0.95428	1.21002	1.47867	1.75288
3	Present	0.44122	0.71206	1.06769	1.27295	1.61076	1.78279
	Ref.	0.44121	0.71206	1.06769	1.27295	1.61076	1.78280
4	Present	0.55006	0.82384	1.18781	1.35816	1.71423	1.89819
	Ref.	0.55006	0.82384	1.18781	1.35817	1.71424	1.89820

5.11 Key Symbols

A	stress amplitude
E	Young's modulus
G, μ	shear modulus, modulus of rigidity
H_i	thickness of the layer i
H	thickness of the cylinder
L	axial half wavelength
n	wave number
$n = 0$	breathing mode
$n = 1$	rigid body mode
$n > 1$	lobar modes
r	radial direction
$r = a$	inner surface boundary
$r = b$	outer surface boundary
R	mean radius of the cylinder
t	time
u_r, u_θ, u_z	displacements in r, θ, z directions
$\mathbf{X}$	vector of constants, vector of unknowns
z	axial direction
$\zeta = \pi/L$	axial wave number
η	loss-factor
θ	transverse direction
λ	Lame's elastic constant
μ, G	shear modulus, modulus of rigidity
ν	Poisson's ratio
ρ	density
σ	normal stress
σ_{rr}, $\sigma_{\theta\theta}$, σ_{zz}	direct stress in the r, θ, z directions
τ	shear stress
$\tau_{r\theta}$	shear stress along θ and perpendicular to r
τ_{rz}	shear stress along z and perpendicular to r
$\tau_{\theta z}$	shear stress along z and perpendicular to θ
ω	excitation frequency
ω_s	the lowest simple thickness shear frequency
$\Omega = \Re\omega/\omega_s$	frequency factor
$\Im\omega$	imaginary part of ω
$\Re\omega$	real part of ω

6

Constrained Layer Damping Treatment of Thick Cylindrical Structures

The provided solution for determination of modal characteristics of a continuous, elastic, single-layer, thick cylinder is extended to a single viscoelastic layer by considering complex elastic moduli for the medium. The systematic solution for each cylindrical layer is formulated to develop a solution for a damped three-layer cylindrical structure. It has been proven that this type of passive damping treatment is an effective vibration suppression technique for most structures. This chapter provids the required mathematical model along with the computed results for a number of case studies.

6.1 Introduction

Damping is a significant dynamic parameter for vibration, sound control, dynamic stability, positioning accuracy, fatigue endurance, and impact resistance. Many current applications of circular cylindrical structures require high dynamic performance. Typical examples of these structures can be found in: aerospace industry, power generation or transmission, heat exchanger tubes, piping systems, nuclear reactor components, smoke stacks, and undersea technology. Therefore, possible sources of passive damping should be compatible with structural configuration.

With the continuing evolution of composite structures and simultaneous requirements for higher performance and lower operational costs, more and more composites are targeted toward structural applications. Composite structures are principally preferred in such applications because of their advanced elastic properties for individual design requirements. They also have the potential for incorporating significant passive damping into the structure.

The potential damping source satisfying the previous requirements is a constrained layer damping treatment. Constrained damping in isotropic

H.R. Hamidzadeh, R.N. Jazar, *Vibrations of Thick Cylindrical Structures*
DOI 10.1007/978-0-387-75591-5_6, © Springer Science+Business Media, LLC 2010

metallic structures has been applied and investigated. The inherent damping capacity of composite structures also seems promising. Although the damping of composite structures is not high, it is significantly higher than most common metallic structures. Moreover, visco-elastic materials are suitable in many cases, as they readily provide high specific stiffness and strength.

Composite structures with visco-elastic damping layers offer the advantages of high damping and good mechanical properties. In addition, the passive damping concept is highly compatible with the laminated configuration of composite structures. The constrained layer damping treatment may produce higher and also adequate damping. However, other critical mechanical properties, such as stiffness and strength, are expected to be reduced. It is desirable to develop integrated composite configurations that can efficiently enhance the damping of composite structures, with higher strength and lower added weight for the structure. Analytical work has been reported in this area, including damping analysis of three-layered cylinders with a constrained layer of visco-elastic material. The effects of different geometries damping layer thickness have also been investigated.

The main objective in using visco-elastic material is to dissipate energy and reduce the amplitude of the vibration. The visco-elastic material can be mathematically modeled using a complex stiffness and elastic moduli. This is due to the phase difference between stress and strain. Visco-elastic materials are very resilient and are widely used for passive damping treatment. The use of such materials reduces the amplitude of stresses and displacements, resulting in a longer working life. The appropriate use of damping material is to sandwich it between two elastic materials. Elastic materials will give the structure stiffness, while damping material will dissipate much of the energy.

The problem of dissipating vibration energy in cylindrical structures to prevent resonant failure has been an important consideration in structural design. Approximate approaches based on curved beam and thin shell theories were reported by the following investigators. Yu (1960) employed shell theory to investigate the axial and torsional vibrations of sandwich cylindrical shells. Pan (1969) studied the damping of circular structures utilizing a visco-elastic core layer. Lu et al. (1973) obtained impedance response of a damping three-layer sandwich ring by adopting the curved beam theory. Lu et al. (1976) formulated the forced vibration of damping ring based on the shell theory. El-Rahed and Wanger (1986) considered the damping response of a shell by using a constrained visco-elastic layer.

Hamidzadeh (1991) studied the damping of lobar vibrations in cylindrical structures. He developed an analytical solution for the three-layer sandwiched thick cylinders with a visco-elastic middle layer by utilizing the stresses and displacements at all interfaces, and by complying with the compatibility requirement at each interface. Hamidzadeh and Chandler (1991) analytically simulated one of the cases presented by Lu et al. (1973) using forced vibration analysis. Their results compared satisfactorily with those of Lu et al. Hamidzadeh and Sawaya (1995) studied the effect of core material loss-factor, and the effect of the outer layer thickness on the natural frequencies and their corresponding modal loss-factor for a three-layered cylinder. Hamidzadeh and Jaing (1995) and Hamidzadeh (2009) employed Hamidzadeh and Sawaya (1995) analytical method and developed a technique to consider the effect of different material loss-factors and thicknesses for constrained layer, as well as different ratios of shear modulus for visco-elastic layer to the elastic one.

6.2 Scope of the Chapter

The constrained layer damping treatment with a visco-elastic core layer for a thick cylindrical structure is considered. An analytical procedure based on the solution that was developed in chapter 5 is employed for the three-layer configuration, which is depicted in Figure 6.1. In connection with the presented analytical method, the modal damping and natural frequencies for long composite circular structures are computed. The effects of geometry on the natural frequencies and loss-factors for different modes of a three-layer sandwiched thick infinitely long cylinder are discussed and presented.

6.3 Natural Frequencies and Modal Loss-Factors

The determinant of the 3×3 coefficient matrix presented by equation (5.38) can provide the natural frequencies of the composite cylinders. Since elements of the matrix are in terms of Bessel functions, the equation will have an infinite number of solutions for any circumferential wave number. An attempt is made to determine only the first six roots of equation (5.27). The dimensionless natural frequency and modal loss-factor for different modes can be computed using Muller's method. This method may be used to find any prescribed number of roots, real or complex, of an arbitrary function. Muller's method (Conte 1965, and Conte and DeBoor 1980) does not require the evaluation of the derivative of the function, and obtains both real and complex roots. As described in section 5.7, natural frequency

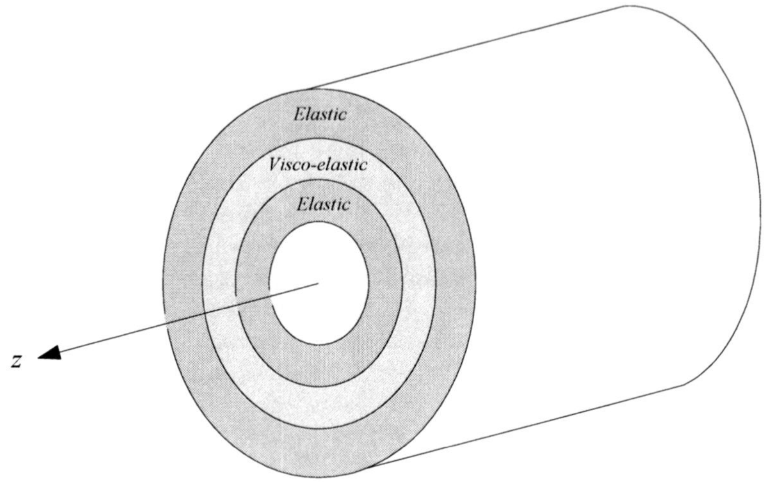

FIGURE 6.1. A typical three-layered sandwich cylindrical structure with visco-elastic middle layer.

factors and the corresponding modal loss-factor are defined by the real and imaginary parts of roots of equation (5.38).

The composite cylinders are considered to be formed by two elastic layers constraining a visco-elastic damping layer. In this chapter, the following three cases are presented and the natural frequency factors and their respective modal loss-factors are computed.

Case 1. In this case, the effects of various material loss-factors for the middle layer on the modal parameters are considered.

Case 2. In this case, the effects of various shear moduli for the middle layer on the modal parameters are considered.

Case 3. In this case, the effects of various thicknesses for the middle layer on the modal parameters are considered.

For each of these cases, the first six dimensionless natural frequency factors and their corresponding loss-factor were computed for five different circumferential wave numbers and different axial wave numbers. Considering that the density of the material is indicated as ρ and ν is Poisson's ratio, and ζ is the axial wave number. Lame's elastic constants, μ and λ, are defined as:

$$\mu = G \tag{6.1}$$

$$\lambda = \frac{2\nu}{1-2\nu}G \tag{6.2}$$

where:

$$G = \frac{E}{2\,(1+\nu)} \tag{6.3}$$

For a linear visco-elastic material, the elastic moduli are presented by complex numbers.

$$E = E'\,(1 + i\eta) \tag{6.4}$$

$$G = G'\,(1 + i\eta) \tag{6.5}$$

E' and G' are the real parts of E and G, respectively. η is defined as the material loss-factor. Considering that the axial wave length ζ is defined as:

$$\zeta = \frac{\pi}{L} \tag{6.6}$$

where L is defined as the half length of the axial wave, the first six roots $\Omega\,(= \omega/\omega_s)$ and their corresponding loss-factors from the frequency equation (5.27) were computed for the value of $H/L = 0.1$, and for a wide range of core shear moduli, core material loss-factors, and thicknesses of visco-elastic layers. The numerical computations were performed for the following material property and range of parameters:

Poisson's ratio: $\nu = 0.3$
Ratio of thickness to mean radius: $0.3 \leq H/R \leq 0.5$
Ratio of thickness to half-wavelength: $H/L = 0.1$
Circumferential waves number: $n = 0, 1, 2, 3, 4$

It should be noted that the accuracy of the computations, i.e. the ratio of the possible error to the value of the root was set to be less than 10^{-5}.

6.4 The Effects of the Core Loss-Factors on Modal Parameters

The effects of various material damping for the middle layer, η_2, on modal parameters are studied. Dimensionless natural frequencies for the first six modes and their corresponding loss-factor η were computed for a three-layer cylinder with different properties. Tables 6.1 to 6.5 show the computed results for a composite cylinder of:

$H/R = 0.3$, $H/L = 0.1$, for $n = 0, 1, 2, 3$, and 4.

The composite cylinder has radii of:

$$
\begin{aligned}
r_0 &= 1.0\,\text{in},\\
r_1 &= 1.15\,\text{in},\\
r_2 &= 1.2\,\text{in, and}\\
r_3 &= 1.35924\,\text{in}.
\end{aligned}
$$

Thicknesses are:

$$
\begin{aligned}
H_1 &= 0.15\,\text{in},\\
H_2 &= 0.05\,\text{in, and}\\
H_3 &= 0.15924\,\text{in}.
\end{aligned}
$$

Layer I and III were assumed to have the same properties as steel with:

$$
\begin{aligned}
G_1 &= 1.1538 \times 10^7\,\text{lb/in}^2,\\
G_2 &= G_2'(1 + i\eta_2),\\
G_2' &= G_1, \text{ and}\\
\nu_1 &= \nu_2 = \nu_3 = 0.3.
\end{aligned}
$$

It is concluded that all natural frequency factors were slightly changed when the material loss-factor for the middle layer (η_2) increased. However, for the circumferential wave number $n = 0$, all six modes of the modal loss-factors increased significantly and linearly with the material loss-factor for the middle layer (η_2) except for the first thickness mode. In case of $n \geq 1$, for the lowest mode, the modal loss-factors varied with different circumferential numbers. It should be noted that the visco-elastic layer would be more efficient for the modes that are associated with large shear deformation.

Table 6.1. Variation of dimensionless natural frequency factors and modal loss-factors for the first six thickness modes with respect to the core loss-factor for a three layer sandwich cylinder.

$n = 0$						
	I		*II*		*III*	
η_2	Ω	η	Ω	η	Ω	η
0.00	0.13877	0.00000	0.18721	0.00000	1.02055	0.00000
0.05	0.13897	0.02563	0.18690	0.00040	1.02228	0.01378
0.10	0.13909	0.03911	0.18666	0.00119	1.02308	0.02736
0.15	0.13969	0.04883	0.18634	0.00222	1.02440	0.04070
0.20	0.14040	0.06016	0.18595	0.00384	1.02622	0.05354
0.25	0.14112	0.07165	0.18550	0.00584	1.02850	0.06580
0.30	0.14183	0.08293	0.18500	0.00825	1.03122	0.07736
0.35	0.14249	0.09393	0.18446	0.01097	1.03431	0.08814
0.40	0.14312	0.10467	0.18390	0.01391	1.03775	0.09808
0.45	0.14372	0.11523	0.18331	0.01695	1.04147	0.10713
0.50	0.14427	0.12569	0.18272	0.02003	1.04543	0.11528

Table 6.1. Continued

$n = 0$						
	IV		*V*		*VI*	
η_2	Ω	η	Ω	η	Ω	η
0.00	1.85680	0.00000	2.01119	0.00000	2.03137	0.00000
0.05	1.85769	0.01021	2.01123	0.00094	2.03156	0.00810
0.10	1.86033	0.02050	2.01137	0.00186	2.03217	0.01636
0.15	1.86457	0.03091	2.01160	0.00275	2.03333	0.02490
0.20	1.87016	0.04145	2.01192	0.00360	2.03525	0.03374
0.25	1.87682	0.05208	2.01230	0.00440	2.03812	0.04277
0.30	1.88424	0.06265	2.01275	0.00514	2.04205	0.05182
0.35	1.89217	0.07296	2.01324	0.00580	2.04710	0.06067
0.40	1.90043	0.08281	2.01379	0.00644	2.05317	0.06908
0.45	1.90889	0.09202	2.01441	0.00690	2.06003	0.07743
0.50	1.91750	0.10046	2.01477	0.00736	2.06783	0.08409

Table 6.2. Variation of dimensionless natural frequency factors and modal loss-factors for the first six thickness modes with respect to the core loss-factor for a three layer sandwich cylinder.

$n = 1$						
	I		II		III	
η_2	Ω	η	Ω	η	Ω	η
0.00	0.08745	0.00000	0.19966	0.00000	0.98492	0.00000
0.05	0.08780	0.12028	0.19969	0.00506	0.97885	0.17430
0.10	0.08846	0.10540	0.19975	0.00979	0.97354	0.21628
0.15	0.08929	0.08767	0.19983	0.01394	0.96978	0.25855
0.20	0.09026	0.06792	0.19990	0.01745	0.96754	0.30079
0.25	0.09141	0.04629	0.19994	0.02032	0.96661	0.34283
0.30	0.09260	0.02655	0.19997	0.02260	0.96674	0.38470
0.35	0.09293	0.01825	0.19987	0.02432	0.96766	0.42661
0.40	0.09277	0.01523	0.19975	0.02545	0.96913	0.46883
0.45	0.09248	0.01373	0.19960	0.02594	0.97092	0.51159
0.50	0.09209	0.01278	0.19940	0.02566	0.97283	0.55506

Table 6.2. Continued

$n = 1$						
	IV		V		VI	
η_2	Ω	η	Ω	η	Ω	η
0.00	1.05804	0.00000	1.57644	0.00000	2.02954	0.00000
0.05	1.05886	0.06209	1.57651	0.00868	2.02928	0.09080
0.10	1.05913	0.08789	1.57654	0.02300	2.02905	0.10166
0.15	1.06213	0.09771	1.57674	0.02822	2.02845	0.10656
0.20	1.06557	0.10071	1.57724	0.03018	2.02779	0.11028
0.25	1.06897	0.10057	1.57782	0.03073	2.02769	0.11393
0.30	1.07224	0.09878	1.57837	0.03063	2.02796	0.11772
0.35	1.07534	0.09604	1.57888	0.03019	2.02880	0.12179
0.40	1.07826	0.09271	1.57934	0.02960	2.03024	0.12597
0.45	1.08099	0.08902	1.57976	0.02893	2.03227	0.13004
0.50	1.08355	0.08508	1.58015	0.02820	2.03544	0.13402

Table 6.3. Variation of dimensionless natural frequency factors and modal loss-factors for the first six thickness modes with respect to the core loss-factor for a three layer sandwich cylinder.

$n = 2$						
	I		*II*		*III*	
η_2	Ω	η	Ω	η	Ω	η
0.00	0.04688	0.00000	0.26172	0.00000	0.93554	0.00000
0.05	0.04709	0.15535	0.26265	0.10354	0.94875	0.16863
0.10	0.04730	0.15334	0.26351	0.10210	0.95435	0.15751
0.15	0.04750	0.15147	0.26430	0.10067	0.96079	0.14785
0.20	0.04769	0.14975	0.26502	0.09926	0.96588	0.13781
0.25	0.04781	0.14796	0.26566	0.09789	0.97037	0.12789
0.30	0.04803	0.14636	0.26629	0.09648	0.97451	0.11847
0.35	0.04819	0.14484	0.26672	0.09512	0.97838	0.10974
0.40	0.04833	0.14341	0.26715	0.09377	0.98205	0.10184
0.45	0.04847	0.14204	0.26751	0.09243	0.98551	0.09483
0.50	0.04860	0.14072	0.26781	0.09109	0.98875	0.08834

Table 6.3. Continued

$n = 2$						
	IV		*V*		*VI*	
η_2	Ω	η	Ω	η	Ω	η
0.00	1.19231	0.00000	1.71527	0.00000	2.22918	0.00000
0.05	1.19448	0.09537	1.71550	0.02889	2.25657	0.01920
0.10	1.19627	0.10105	1.71578	0.04263	2.25399	0.01543
0.15	1.19799	0.10234	1.71636	0.04741	2.25226	0.01408
0.20	1.19986	0.10173	1.71738	0.04892	2.25152	0.01333
0.25	1.20179	0.10017	1.71844	0.04906	2.25116	0.01275
0.30	1.20373	0.09812	1.71941	0.04852	2.25097	0.01223
0.35	1.20561	0.09579	1.72027	0.04764	2.25087	0.01176
0.40	1.20741	0.09333	1.72101	0.04660	2.25084	0.01133
0.45	1.20911	0.09083	1.72166	0.04538	2.25084	0.01094
0.50	1.21070	0.08834	1.72223	0.04412	2.25087	0.01058

Table 6.4. Variation of dimensionless natural frequency factors and modal loss-factors for the first six thickness modes with respect to the core loss-factor for a three layer sandwich cylinder.

$n = 3$						
	I		II		III	
η_2	Ω	η	Ω	η	Ω	η
0.00	0.29156	0.00000	0.80541	0.00000	1.00639	0.00000
0.05	0.29144	0.06309	0.81238	0.01993	1.00081	0.21529
0.10	0.29103	0.07327	0.81831	0.01861	0.99611	0.25034
0.15	0.29174	0.07744	0.82277	0.01729	0.99237	0.28485
0.20	0.29229	0.07466	0.82607	0.01571	0.98958	0.32208
0.25	0.29332	0.07535	0.82845	0.01420	0.98773	0.35889
0.30	0.29469	0.07952	0.83016	0.01275	0.98676	0.39601
0.35	0.29627	0.08560	0.83138	0.01134	0.98661	0.43309
0.40	0.29798	0.09257	0.83225	0.00995	0.98722	0.46975
0.45	0.29969	0.09992	0.83287	0.00859	0.98856	0.50547
0.50	0.30144	0.10739	0.83330	0.00728	0.99068	0.53954

Table 6.4. Continued

$n = 3$						
	IV		V		VI	
η_2	Ω	η	Ω	η	Ω	η
0.00	1.51690	0.00000	1.85923	0.00000	2.03437	0.00000
0.05	1.52147	0.02446	1.85617	0.05249	2.03441	0.10668
0.10	1.52579	0.02337	1.85392	0.06672	2.03474	0.11612
0.15	1.52871	0.02207	1.85366	0.07112	2.03535	0.12134
0.20	1.53011	0.02099	1.85542	0.07223	2.03629	0.12553
0.25	1.53096	0.02050	1.85737	0.07188	2.03767	0.12844
0.30	1.53157	0.02035	1.85924	0.07076	2.03957	0.13325
0.35	1.53207	0.02038	1.86090	0.06919	2.04204	0.13692
0.40	1.53250	0.02051	1.86236	0.06733	2.04498	0.14023
0.45	1.53290	0.02068	1.86358	0.06527	2.04849	0.14323
0.50	1.53327	0.02088	1.86452	0.06308	2.05194	0.14646

Table 6.5. Variation of dimensionless natural frequency factors and modal loss-factors for the first six thickness modes with respect to the core loss-factor for a three layer sandwich cylinder.

$n = 4$

	I		*II*		*III*	
η_2	Ω	η	Ω	η	Ω	η
0.00	0.38209	0.00000	0.60055	0.00000	1.07525	0.00000
0.05	0.38222	0.00702	0.60077	0.00350	1.06851	0.04309
0.10	0.38260	0.01403	0.60135	0.00691	1.06480	0.02902
0.15	0.38320	0.02105	0.60211	0.01029	1.06327	0.02223
0.20	0.38400	0.02805	0.60289	0.01328	1.06257	0.01840
0.25	0.38496	0.03501	0.60356	0.01593	1.06223	0.01595
0.30	0.38605	0.04193	0.60408	0.01838	1.06209	0.01423
0.35	0.38723	0.04877	0.60443	0.02082	1.06201	0.01293
0.40	0.38847	0.05554	0.60462	0.02343	1.06217	0.01192
0.45	0.38974	0.06221	0.60471	0.02641	1.06233	0.01112
0.50	0.39102	0.06877	0.60476	0.02889	1.06254	0.01050

Table 6.5. Continued

$n = 4$

	IV		*V*		*VI*	
η_2	Ω	η	Ω	η	Ω	η
0.00	1.43243	0.00000	1.75467	0.00000	2.04933	0.00000
0.05	1.43050	0.07054	1.75477	0.01231	2.05866	0.01990
0.10	1.42939	0.07759	1.75494	0.01210	2.06683	0.01771
0.15	1.42992	0.07588	1.75553	0.01128	2.06920	0.01589
0.20	1.43152	0.07419	1.75605	0.01038	2.07014	0.01412
0.25	1.43363	0.07261	1.75605	0.00953	2.07070	0.01326
0.30	1.43590	0.07119	1.75586	0.00874	2.07116	0.01212
0.35	1.43816	0.06992	1.75559	0.00799	2.07161	0.01103
0.40	1.44031	0.06876	1.75532	0.00731	2.07205	0.00995
0.45	1.44233	0.06770	1.75506	0.00666	2.07247	0.00889
0.50	1.44419	0.00667	1.75482	0.00606	2.07286	0.00786

Using the data of Tables 6.1-6.5, figures 6.2 to 6.6 depict the natural frequency factor Ω as a function of material loss factor η_2 for the first three modes of $n = 0$, $n = 1$, $n = 2$, $n = 3$, $n = 4$. Figures 6.7 to 6.11 show the modal loss factor η as a function of material loss factor η_2 for the first three modes of $n = 0$, $n = 1$, $n = 2$, $n = 3$, $n = 4$.

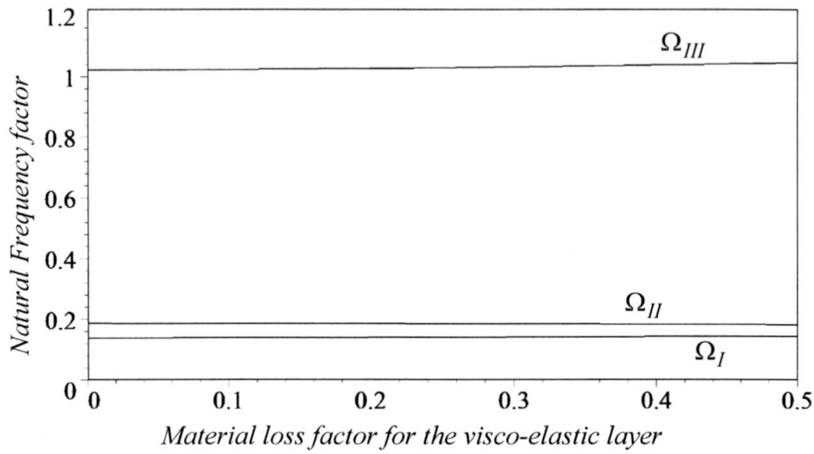

FIGURE 6.2. Variation of the natural frequency factors for the first three modes versus the material loss-factor of the viscoelastic layer for $n = 0$.

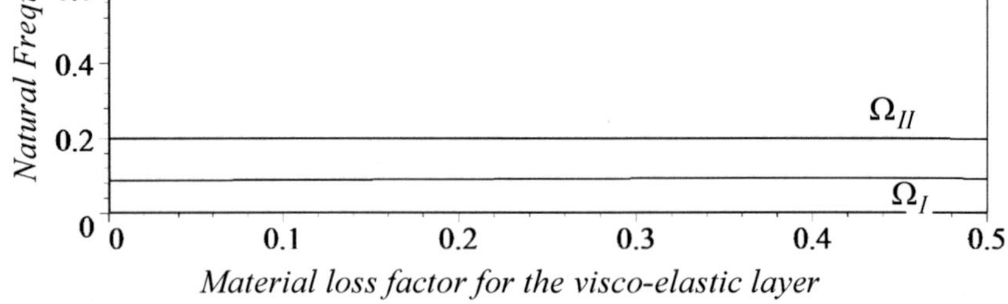

FIGURE 6.3. Variation of the natural frequency factors for the first three modes versus the material loss-factor of the viscoelastic layer for $n = 1$.

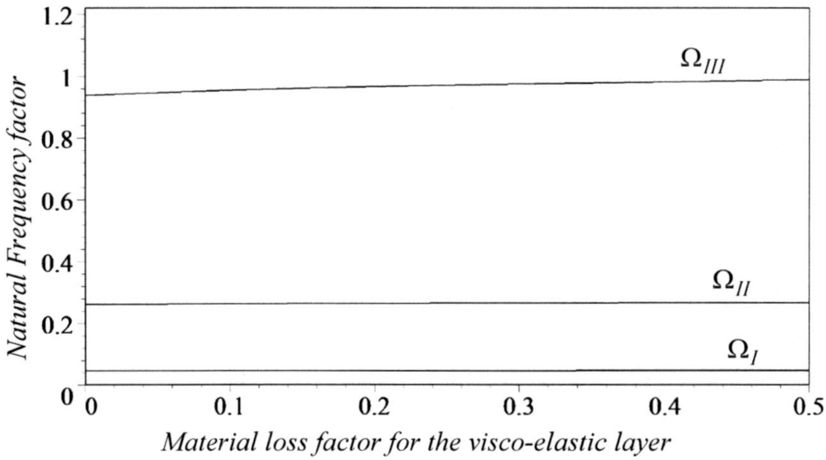

FIGURE 6.4. Variation of the natural frequency factors for the first three modes versus the material loss-factor of the viscoelastic layer for $n = 2$.

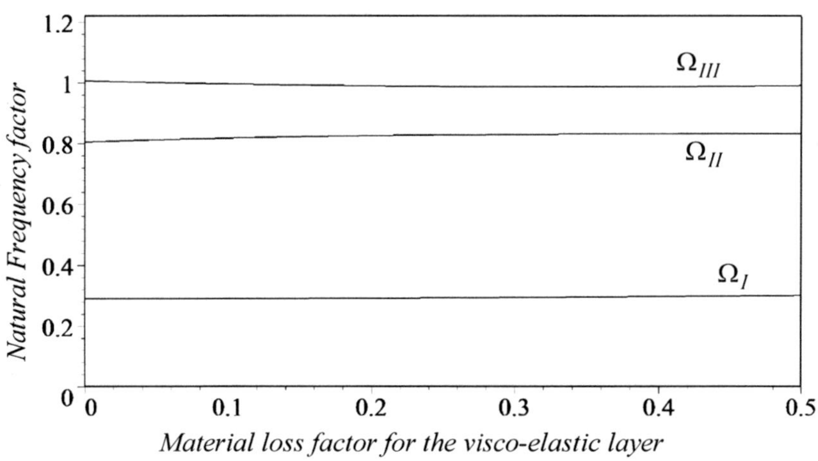

FIGURE 6.5. Variation of the natural frequency factors for the first three modes versus the material loss-factor of the viscoelastic layer for $n = 3$.

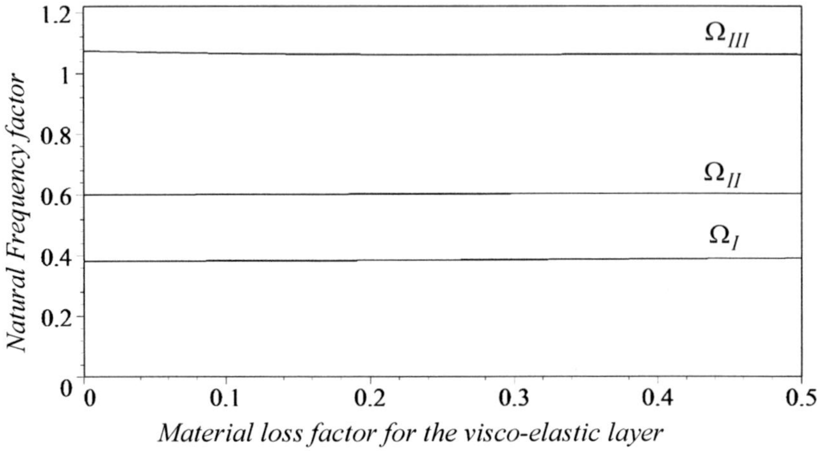

FIGURE 6.6. Variation of the natural frequency factors for the first three modes versus the material loss-factor of the viscoelastic layer for $n = 4$.

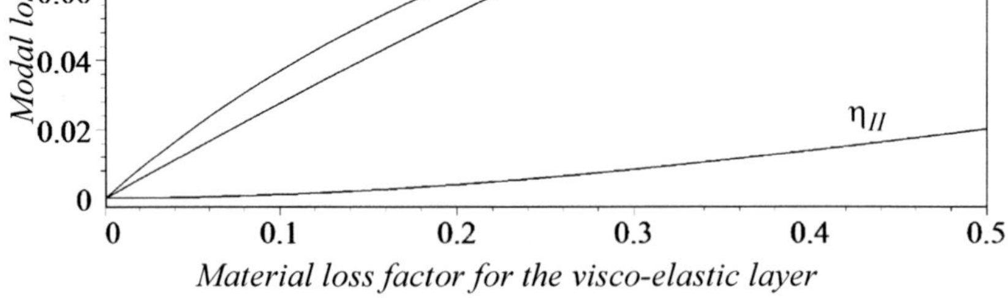

FIGURE 6.7. Variation of the modal loss factors for the first three modes versus the material loss-factor of the viscoelastic layer for $n = 0$.

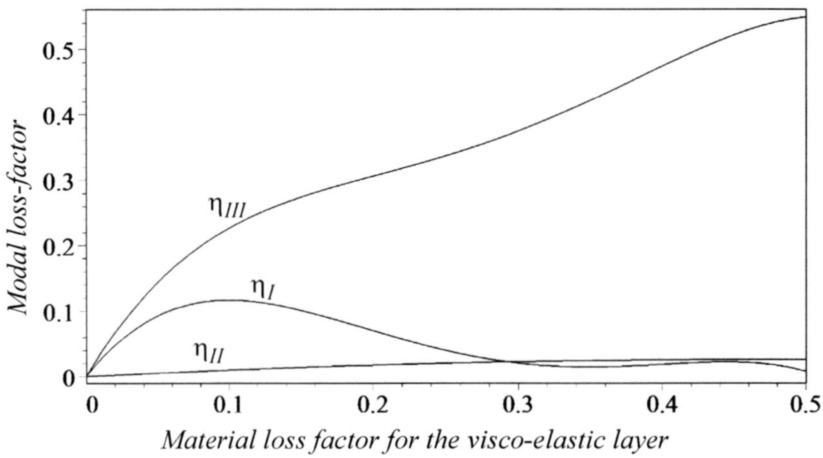

FIGURE 6.8. Variation of the modal loss factors for the first three modes versus the material loss-factor of the viscoelastic layer for $n = 1$.

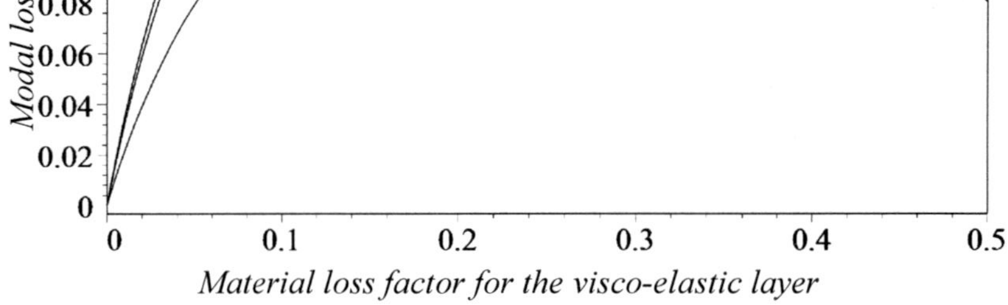

FIGURE 6.9. Variation of the modal loss factors for the first three modes versus the material loss-factor of the viscoelastic layer for $n = 2$.

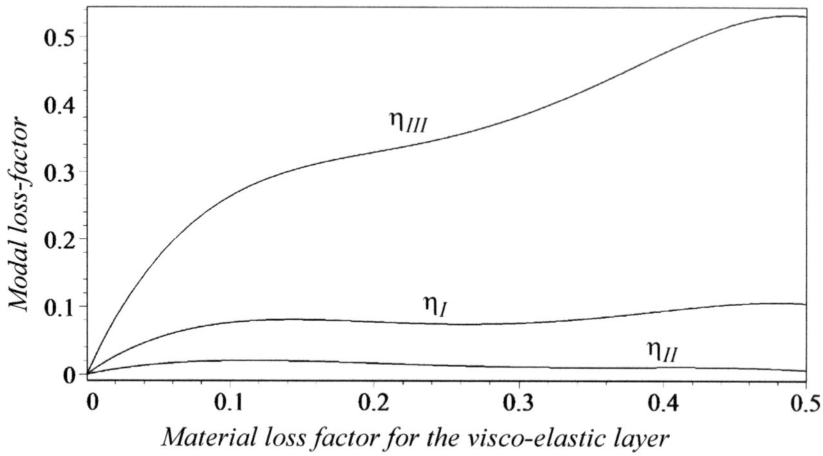

FIGURE 6.10. Variation of the modal loss factors for the first three modes versus the material loss-factor of the viscoelastic layer for $n = 3$.

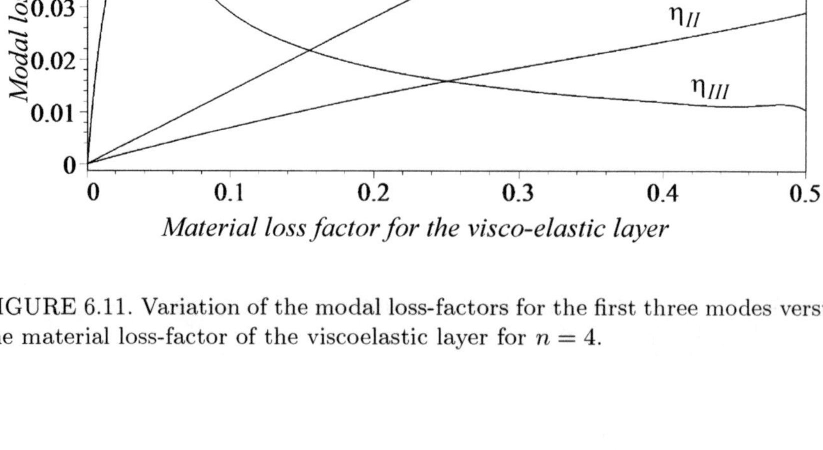

FIGURE 6.11. Variation of the modal loss-factors for the first three modes versus the material loss-factor of the viscoelastic layer for $n = 4$.

6.5 The Effects of Various Shear Moduli for the Middle Layer

Dimensionless natural frequency factors, Ω, for the first six modes and their corresponding loss-factors, η, are computed for a three-layer fully-covered sandwiched cylinder with the middle layer having a loss-factor of $\eta_2 = 0.05$. The shear modulus of the middle layer is varied. Tables 6.6-6.10 show the results obtained for $H/R = 0.3$, $H/L = 0.1$, at $n = 0, 1, 2, 3$, and 4 for a laminated cylinder with radii of:

$r_0 = 1.0\,\text{in}$,
$r_1 = 1.15\,\text{in}$,
$r_2 = 1.2\,\text{in}$, and
$r_3 = 1.35294\,\text{in}$.

Thicknesses are:

$H_1 = 0.15\,\text{in}$,
$H_2 = 0.05\,\text{in}$, and
$H_3 = 0.15924\,\text{in}$.

Layer I and III are assumed to have the same elastic properties with

$G_1 = 1.1538 \times 10^7\,\text{lb/in}^2$.
$\nu_1 = \nu_2 = \nu_3 = 0.3$

It is concluded that for all reported values of the circumferential wave numbers, the natural frequency factors for all thickness modes increased slightly when the core shear modulus was increased. This is due to the fact that overall stiffness in the whole structure increases as the ratio of G_2'/G_1 is increased.

Using the data of Tables 6.6 to 6.10, figures 6.12 to 6.16 depict the natural frequency factor Ω as a function of G_2'/G_1 for the first three modes of $n = 0$, $n = 1$, $n = 2$, $n = 3$, $n = 4$. Figures 6.17 to 6.21 show the modal loss factor η as a function of G_2'/G_1 for the first three modes of $n = 0$, $n = 1$, $n = 2$, $n = 3$, $n = 4$.

Table 6.6. Variation of dimensionless frequency factors and modal loss-factors for the first six modes with respect to the ratio of G_2'/G_1.

$n = 0$						
	I		*II*		*III*	
G_2'/G_1	Ω	η	Ω	η	Ω	η
1.000	0.13877	0.03563	0.18944	0.00705	1.02228	0.01378
0.950	0.13845	0.04389	0.18877	0.00659	1.01492	0.01436
0.925	0.13834	0.04675	0.18844	0.00635	1.01100	0.01466
0.900	0.13819	0.04922	0.18811	0.00612	1.00690	0.01498
0.875	0.13801	0.05138	0.18792	0.00586	1.00072	0.01523
0.850	0.13787	0.05346	0.18745	0.00566	0.99811	0.01566
0.825	0.13770	0.05535	0.18711	0.00542	0.99340	0.01602
0.800	0.13754	0.05713	0.18708	0.00272	0.98845	0.01640
0.780	0.13741	0.05848	0.18649	0.00022	0.98431	0.01671
0.760	0.13727	0.05978	0.18862	0.00030	0.98000	0.01704
0.750	0.13721	0.06041	0.18609	0.00034	0.97778	0.01720
0.740	0.13714	0.06103	0.18596	0.00037	0.97551	0.01737
0.730	0.13707	0.06163	0.18583	0.00040	0.97319	0.01754
0.720	0.13701	0.06222	0.18569	0.00043	0.97082	0.01772
0.710	0.13694	0.06280	0.18556	0.00456	0.96840	0.01790
0.700	0.13688	0.06336	0.18543	0.00048	0.96593	0.01809
0.675	0.13670	0.06467	0.18509	0.00053	0.95951	0.01856
0.650	0.13653	0.06585	0.18471	0.00059	0.95271	0.01906
0.625	0.13634	0.06679	0.18441	0.00059	0.94551	0.01958
0.600	0.13614	0.06735	0.18407	0.00061	0.93787	0.02013
0.575	0.13594	0.06811	0.18373	0.00061	0.92974	0.02071
0.550	0.13584	0.07093	0.18339	0.00060	0.92109	0.02132
0.525	0.13579	0.07512	0.18305	0.00058	0.91185.	0.02196
0.500	0.13575	0.07831	0.18275	0.00056	0.90198	0.02263

Table 6.6. Continued

$n = 0$						
	IV		V		VI	
G_2'/G_1	Ω	η	Ω	η	Ω	η
1.000	1.85769	0.01021	2.01123	0.00094	2.03156	0.00810
0.950	1.84567	0.01109	2.01024	0.00099	2.02901	0.00839
0.925	1.83918	0.01154	2.00970	0.00102	2.02771	0.00852
0.900	1.83235	0.01200	2.00914	0.00105	2.02640	0.00865
0.875	1.82564	0.01254	2.00823	0.00108	2.02514	0.00874
0.850	1.81758	0.01294	2.00790	0.00111	2.02372	0.00888
0.825	1.80959	0.01343	2.00722	0.00115	2.02234	0.00899
0.800	1.80115	0.01393	2.00650	0.00118	2.02093	0.00909
0.780	1.79406	0.01434	2.00589	0.00122	2.01978	0.00917
0.760	1.78664	0.01474	2.00525	0.00125	2.01861	0.00924
0.750	1.78280	0.01496	2.00491	0.00127	2.01808	0.00927
0.740	1.77888	0.01518	2.00457	0.00129	2.01741	0.00930
0.730	1.77486	0.01539	2.00421	0.00131	2.01679	0.00933
0.720	1.77076	0.01561	2.00385	0.00133	2.01617	0.00936
0.710	1.76655	0.01583	2.00347	0.00135	2.01555	0.00939
0.700	1.76224	0.01606	2.00309	0.00137	2.01491	0.00942
0.675	1.75102	0.01663	2.00207	0.00143	2.01327	0.00948
0.650	1.73910	0.01906	2.00096	0.00149	2.01156	0.00954
0.625	1.72643	0.01783	1.99977	0.00156	2.00978	0.00958
0.600	1.71293	0.01846	1.99847	0.00163	2.00790	0.00961
0.575	1.69852	0.01912	1.99705	0.00171	2.00591	0.00963
0.550	1.68314	0.01981	1.99549	0.00180	2.00380	0.00965
0.525	1.66667	0.02052	1.99377	0.00190	2.00154	0.00964
0.500	1.64900	0.02126	1.99186	0.00202	1.99991	0.00963

Table 6.7. Variation of dimensionless frequency factors and modal loss-factors for the first six modes with respect to the ratio of G_2'/G_1.

$n = 1$						
	I		II		III	
G_2'/G_1	Ω	η	Ω	η	Ω	η
1.000	0.08780	0.12028	0.19969	0.00506	0.97885	0.17430
0.950	0.08779	0.11162	0.19914	0.00542	0.95943	0.17116
0.925	0.08788	0.10339	0.19885	0.00574	0.94881	0.16906
0.900	0.08861	0.08974	0.19853	0.00622	0.93719	0.16699
0.875	0.09101	0.08384	0.19817	0.00689	0.92426	0.16611
0.850	0.09028	0.04689	0.19776	0.00781	0.91920	0.16856
0.825	0.08933	0.04089	0.19727	0.00925	0.89609	0.17578
0.800	0.08933	0.03585	0.19669	0.01125	0.88311	0.18645
0.780	0.09111	0.04012	0.19610	0.01346	0.87370	0.19600
0.760	0.08718	0.04999	0.19537	0.01634	0.86501	0.20553
0.750	0.08666	0.05000	0.19493	0.01805	0.86090	0.21018
0.740	0.08602	0.05000	0.19444	0.01991	0.85691	0.21470
0.730	0.08503	0.04801	0.19387	0.02188	0.85303	0.21908
0.720	0.08431	0.02356	0.19324	0.02389	0.84926	0.22329
0.710	0.08678	0.01169	0.19254	0.02587	0.84558	0.22733
0.700	0.08677	0.01523	0.19178	0.02774	0.84198	0.23117
0.675	0.08574	0.02744	0.18962	0.03168	0.83333	0.23978
0.650	0.08062	0.05000	0.19423	0.01086	0.82507	0.24678
0.625	0.08766	0.05901	0.20021	0.00717	0.81717	0.25189
0.600	0.07745	0.05000	0.19960	0.00611	0.80959	0.25476
0.575	0.06667	0.05742	0.19903	0.00553	0.80231	0.25493
0.550	0.06626	0.05546	0.19847	0.00530	0.79539	0.25166
0.525	0.06575	0.05392	0.19790	0.00534	0.78885	0.24361
0.500	0.06510	0.05274	0.19729	0.00563	0.78246	0.22833

Table 6.7. Continued

$n = 1$						
	IV		V		VI	
G_2'/G_1	Ω	η	Ω	η	Ω	η
1.000	1.05986	0.06209	1.57651	0.00868	1.94834	0.12904
0.950	1.07072	0.08169	1.59151	0.02068	1.90487	0.13455
0.925	1.07150	0.09311	1.59295	0.03098	1.88671	0.13817
0.900	1.07028	0.10330	1.59047	0.04044	1.87033	0.14026
0.875	1.06770	0.11189	1.58556	0.04726	1.88551	0.14019
0.850	1.06420	0.11884	1.57952	0.05110	1.84052	0.13746
0.825	1.06003	0.12414	1.57315	0.05189	1.82618	0.13135
0.800	1.05537	0.12776	1.56714	0.04941	1.81130	0.12065
0.780	1.05136	0.12935	1.56320	0.04480	1.96197	0.12125
0.760	1.04717	0.12967	1.56094	0.03762	1.93547	0.12030
0.750	1.04503	0.12929	1.56090	0.03314	1.92161	0.12282
0.740	1.04287	0.12851	1.56190	0.02835	1.90886	0.12689
0.730	1.04070	0.12730	1.56411	0.02369	1.89759	0.13149
0.720	1.03854	0.12558	1.56755	0.01959	1.88673	0.13588
0.710	1.03640	0.12330	1.57245	0.01653	1.87865	0.13973
0.700	1.03434	0.12035	1.57894	0.01616	1.87037	0.14289
0.675	1.02998	0.10918	1.58921	0.02967	1.85162	0.14761
0.650	1.02943	0.09175	1.58747	0.04432	1.83461	0.14748
0.625	1.02998	0.10918	1.58921	0.02967	1.85162	0.14761
0.600	1.04347	0.07477	1.57291	0.05579	1.80339	0.12569
0.575	1.04866	0.08492	1.56599	0.05225	1.88112	0.11689
0.550	1.04729	0.09760	1.56369	0.04174	1.85299	0.14110
0.525	1.04212	0.10617	1.57504	0.03133	1.83236	0.15153
0.500	1.03514	0.10931	1.58473	0.04773	1.81380	0.15230

Table 6.8. Variation of dimensionless frequency factors and modal loss-factors for the first six modes with respect to the ratio of G_2'/G_1.

$n=2$						
	I		II		III	
G_2'/G_1	Ω	η	Ω	η	Ω	η
1.000	0.047096	0.15532	0.21817	0.20558	0.26265	0.10354
0.950	0.046958	0.15530	0.21752	0.20574	0.26164	0.10368
0.925	0.046889	0.15527	0.21718	0.20585	0.26113	0.10375
0.900	0.046820	0.15524	0.21685	0.20598	0.26064	0.10383
0.875	0.046751	0.15520	0.21650	0.20611	0.26014	0.10392
0.850	0.046682	0.15515	0.21615	0.20626	0.25965	0.10402
0.825	0.046611	0.15510	0.21579	0.20641	0.25917	0.10412
0.800	0.046539	0.15504	0.21541	0.20657	0.25869	0.10423
0.780	0.046480	0.15499	0.21509	0.02066	0.25831	0.10432
0.760	0.046419	0.15494	0.21477	0.20679	0.25794	0.10441
0.750	0.046389	0.15491	0.21460	0.20683	0.25776	0.10446
0.740	0.046357	0.15489	0.21443	0.20688	0.25757	0.10451
0.730	0.046326	0.15487	0.21425	0.20691	0.25739	0.10456
0.720	0.046293	0.15485	0.21407	0.20694	0.25721	0.10462
0.710	0.046260	0.15483	0.21388	0.20696	0.25703	0.10467
0.700	0.046227	0.15483	0.21369	0.20697	0.25686	0.10473
0.675	0.046138	0.15480	0.21318	0.20693	0.25642	0.10488
0.650	0.046045	0.15479	0.21262	0.20677	0.25599	0.10504
0.625	0.045945	0.15482	0.21199	0.20645	0.25558	0.10521
0.600	0.045838	0.15492	0.21130	0.20589	0.25518	0.10540
0.575	0.045721	0.15503	0.21049	0.20497	0.25478	0.10561
0.550	0.045595	0.15523	0.20957	0.20358	0.25439	0.10584
0.525	0.045457	0.15552	0.20847	0.20150	0.25401	0.10610
0.500	0.045309	0.15591	0.20717	0.19848	0.25363	0.10640

Table 6.8. Continued

$n = 2$						
	IV		V		VI	
G_2'/G_1	Ω	η	Ω	η	Ω	η
1.000	1.01371	0.08315	1.19448	0.09537	1.71650	0.02889
0.950	0.98830	0.13642	1.19947	0.09860	1.73126	0.04352
0.925	0.98029	0.16375	1.19996	0.10375	1.73081	0.05415
0.900	0.97441	0.18672	1.19886	0.10870	1.72686	0.06242
0.875	0.96928	0.20507	1.19659	0.11302	1.72127	0.06772
0.850	0.96419	0.21987	1.19357	0.11616	1.71513	0.07007
0.825	0.95889	0.23202	1.19013	0.11792	1.70925	0.06942
0.800	0.95333	0.24212	1.18653	0.11822	1.70451	0.06562
0.780	0.94868	0.24893	1.18337	0.11737	1.70248	0.06016
0.760	0.94386	0.25473	1.18108	0.11556	1.70345	0.05292
0.750	0.94140	0.25728	1.17989	0.11431	1.70568	0.04916
0.740	0.93889	0.25959	1.17882	0.11284	1.70934	0.04606
0.730	0.93635	0.26168	1.17786	0.11119	1.71430	0.04457
0.720	0.93377	0.26353	1.17706	0.10938	1.71976	0.04568
0.710	0.93116	0.26516	1.17641	0.10745	1.72423	0.04949
0.700	0.92852	0.26655	1.17593	0.10547	1.72672	0.05483
0.675	0.92176	0.26900	1.17549	0.10048	1.72590	0.06741
0.650	0.91482	0.26987	1.17618	0.09617	1.72023	0.07542
0.625	0.90770	0.26899	1.17790	0.09418	1.71344	0.07827
0.600	0.90045	0.26609	1.17895	0.09658	1.70800	0.07600
0.575	0.89307	0.26076	1.17712	0.10113	1.70743	0.06916
0.550	0.88556	0.25227	1.17283	0.10386	1.71648	0.06599
0.525	0.87788	0.23956	1.16754	0.10281	1.72135	0.07750
0.500	0.86966	0.22097	1.16287	0.09726	1.71655	0.08722

Table 6.9. Variation of dimensionless frequency factors and modal loss-factors for the first six modes with respect to the ratio of G_2'/G_1.

$n = 3$						
	I		II		III	
G_2'/G_1	Ω	η	Ω	η	Ω	η
1.000	0.29156	0.07924	0.68874	0.00928	1.00081	0.21529
0.950	0.29084	0.08276	0.69294	0.01359	0.98642	0.20825
0.925	0.29045	0.08446	0.69526	0.01868	0.97881	0.20408
0.900	0.29016	0.08612	0.69627	0.02670	0.97104	0.19934
0.875	0.28964	0.08774	0.69627	0.03510	0.96312	0.19365
0.850	0.28920	0.08940	0.69437	0.04106	0.95501	0.18657
0.825	0.28874	0.09101	0.69178	0.04382	0.94655	0.17757
0.800	0.28825	0.09261	0.68905	0.04321	0.93746	0.16600
0.780	0.28783	0.09388	0.68707	0.04014	0.92927	0.15439
0.760	0.28738	0.09513	0.68566	0.03473	0.91953	0.14059
0.750	0.28715	0.09576	0.68535	0.03125	0.91372	0.13320
0.740	0.28691	0.09638	0.68539	0.02751	0.90707	0.12611
0.730	0.28666	0.09700	0.68584	0.02383	0.89960	0.12016
0.720	0.28640	0.09761	0.68665	0.02060	0.89162	0.11619
0.710	0.28613	0.09821	0.68772	0.01803	0.88363	0.11441
0.700	0.28585	0.09881	0.68893	0.01613	0.87607	0.11433
0.675	0.28509	0.10025	0.69214	0.01354	0.85972	0.11670
0.650	0.28424	0.10161	0.69546	0.01278	0.84619	0.11733
0.625	0.28326	0.10282	0.69905	0.01293	0.83445	0.11329
0.600	0.28214	0.10382	0.70367	0.01305	0.82448	0.10269
0.575	0.28087	0.10451	0.70616	0.01884	0.85812	0.11438
0.550	0.27927	0.10478	0.71745	0.02757	0.84736	0.11063
0.525	0.27739	0.10443	0.72415	0.03051	0.83642	0.10337
0.500	0.27512	0.10332	0.72272	0.01456	0.82491	0.09101

Table 6.9. Continued

$n = 3$						
	IV		V		VI	
G_2'/G_1	Ω	η	Ω	η	Ω	η
1.000	1.38688	0.07046	1.85617	0.05249	2.03441	0.10668
0.950	1.42820	0.08357	1.75312	0.00466	2.06330	0.01478
0.925	1.33509	0.11916	1.86256	0.08366	2.03898	0.11222
0.900	1.42190	0.08940	1.75883	0.00564	2.07082	0.01046
0.875	1.31468	0.13341	1.85039	0.09340	2.03154	0.11586
0.850	1.30496	0.13551	1.84440	0.09403	2.02665	0.11525
0.825	1.29527	0.13455	1.83956	0.09199	2.02196	0.11256
0.800	1.28546	0.13017	1.83706	0.08738	2.01859	0.10754
0.780	1.27737	0.12357	1.83814	0.08235	2.01813	0.10208
0.760	1.26874	0.11316	1.84435	0.07834	2.02079	0.09702
0.750	1.26399	0.10601	1.84759	0.07832	2.02302	0.09562
0.740	1.25870	0.09712	1.85152	0.08043	2.02521	0.09531
0.730	1.25261	0.08601	1.85415	0.08426	2.02685	0.09592
0.720	1.24558	0.07110	1.85511	0.08877	2.02769	0.09705
0.710	1.24297	0.05075	1.85468	0.09317	2.02772	0.09833
0.700	1.24293	0.04022	1.85327	0.09708	2.02705	0.09951
0.675	1.23643	0.03014	1.84754	0.10401	2.02335	0.10088
0.650	1.26694	0.02976	1.84103	0.10693	2.01852	0.09896
0.625	1.28702	0.09174	1.83582	0.10621	2.01520	0.09296
0.600	1.30347	0.11743	1.83420	0.10301	2.01756	0.08454
0.575	1.28678	0.12494	1.83756	0.10229	2.02491	0.08245
0.550	1.26964	0.12543	1.83882	0.10867	2.02607	0.08571
0.525	1.25118	0.11921	1.83425	0.11500	1.02238	0.08554
0.500	1.22967	0.10439	1.82795	0.11729	2.02005	0.07916

Table 6.10. Variation of dimensionless frequency factors and modal loss-factors for the first six modes with respect to the ratio of G_2'/G_1.

$n = 4$						
	I		*II*		*III*	
G_2'/G_1	Ω	η	Ω	η	Ω	η
1.000	0.38222	0.00702	0.60077	0.00350	1.06851	0.04309
0.950	0.38093	0.00806	0.59948	0.46955	1.05659	0.04662
0.925	0.38024	0.00867	0.59867	0.00540	1.05088	0.04085
0.900	0.37954	0.00937	0.59774	0.00624	1.04757	0.02985
0.875	0.37880	0.01014	0.59663	0.00730	0.97346	0.06948
0.850	0.37803	0.01102	0.59532	0.00867	0.97283	0.08831
0.825	0.37721	0.01201	0.59370	0.01045	0.96949	0.10243
0.800	0.37634	0.01314	0.59169	0.01280	0.96498	0.11260
0.780	0.37560	0.01416	0.58969	0.01520	0.96094	0.11838
0.760	0.37481	0.01529	0.58726	0.01813	0.95667	0.12232
0.750	0.37440	0.01591	0.58584	0.01982	0.96448	0.12362
0.740	0.37397	0.01656	0.58425	0.02163	0.95226	0.12447
0.730	0.37352	0.01725	0.58249	0.02358	0.95002	0.12487
0.720	0.37305	0.01798	0.58055	0.02564	0.94777	0.12479
0.710	0.37258	0.01875	0.57841	0.02779	0.94551	0.12421
0.700	0.37209	0.01956	0.57606	0.03000	0.94327	0.12310
0.675	0.37073	0.02181	0.56922	0.03559	0.93777	0.11763
0.650	0.36921	0.02439	0.56109	0.02439	0.93275	0.10723
0.625	0.36746	0.02732	0.62736	0.00965	0.92904	0.08934
0.600	0.36545	0.03060	0.62540	0.00642	0.92896	0.05895
0.575	0.36309	0.03418	0.62412	0.00350	0.93196	0.02065
0.550	0.36031	0.03796	0.62349	0.00017	0.93376	0.00644
0.525	0.35700	0.04175	0.62341	0.05598	0.93235	0.01248
0.500	0.35305	0.04532	0.62210	0.01551	0.92476	0.02885

Table 6.10. Continued

$n = 4$						
	IV		*V*		*VI*	
G_2'/G_1	Ω	η	Ω	η	Ω	η
1.000	1.25700	0.00478	1.74774	0.01231	2.05866	0.01990
0.950	1.42820	0.08356	1.75312	0.00466	2.06330	0.01052
0.925	1.42567	0.08659	1.75630	0.00477	2.06705	0.00954
0.900	1.42490	0.08939	1.75883	0.00564	2.07082	0.01046
0.875	1.41706	0.09127	1.76134	0.00729	2.07395	0.01377
0.850	1.41145	0.09175	1.76372	0.01056	2.07424	0.01948
0.825	1.40540	0.09049	1.76445	0.01606	2.06989	0.02490
0.800	1.39929	0.08718	1.76171	0.02176	2.06190	0.02651
0.780	1.39467	0.08281	1.75758	0.24335	2.05584	0.02308
0.760	1.39070	0.07671	1.75294	0.24473	2.05396	0.01686
0.750	1.38912	0.07297	1.75076	0.02355	2.05479	0.01415
0.740	1.38792	0.06877	1.74888	0.02195	2.05627	0.01215
0.730	1.38722	0.06416	1.74751	0.01977	2.05805	0.01087
0.720	1.38713	0.05920	1.74687	0.01798	2.05994	0.01023
0.710	1.38779	0.05403	1.74708	0.01458	2.06185	0.01018
0.700	1.38937	0.04880	1.74807	0.01238	2.06370	0.01075
0.675	1.41918	0.04999	1.75213	0.00977	2.06669	0.01510
0.650	1.40799	0.06181	1.75664	0.01097	2.06357	0.02117
0.625	1.40194	0.07204	1.75971	0.016652.	05559	0.02203
0.600	1.39472	0.07862	1.75719	0.02416	2.05091	0.01546
0.575	1.38646	0.08128	1.75055	0.02706	2.05444	0.01010
0.550	1.37814	0.02354	1.74456	0.02354	2.05936	0.01106
0.525	1.37151	0.07096	1.74454	0.01618	2.05896	0.01692
0.500	1.37043	0.05716	1.75054	0.01489	2.05137	0.02160

6.6 The Effects of Various Thicknesses for The Middle Layer

Dimensionless natural frequency factors (Ω) for the first six modes and their corresponding loss-factors (η) are computed for a three-layer, sandwiched cylinder with the visco-elastic core of $\eta_2 = 0.05$. The thickness of the middle layer varies. Tables 6.11 to 6.15 show the results obtained for a composite cylinder of $H/L = 0.1$, for $n = 0, 1, 2, 3$, and 4. The composite cylinder has radii of:

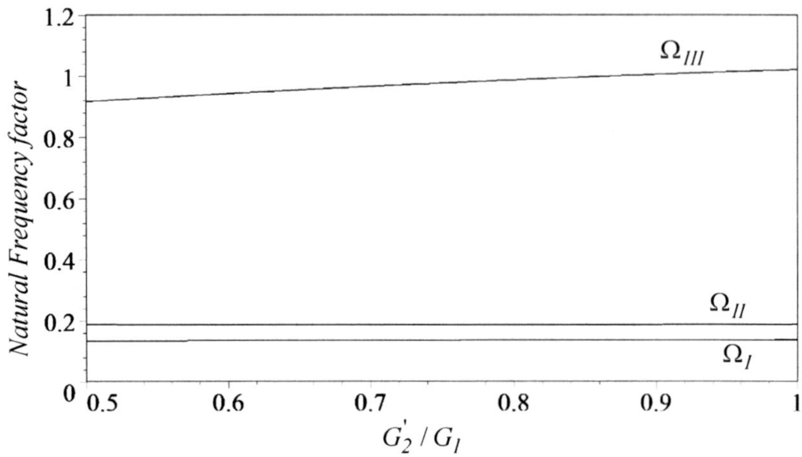

FIGURE 6.12. Variation of natural frequency factors for the first three modes versus ratio of G_2'/G_1 for $n = 0$.

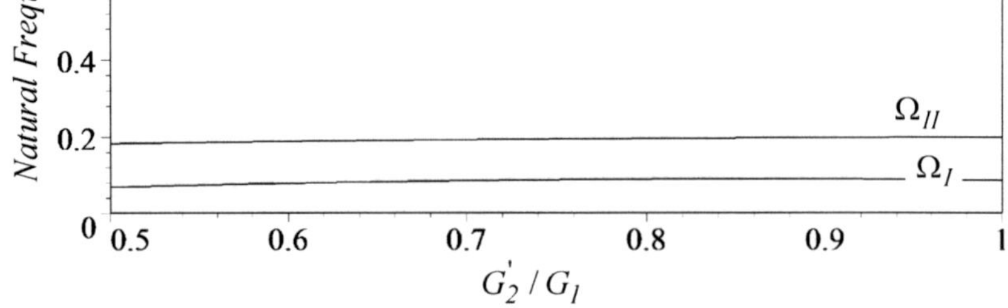

FIGURE 6.13. Variation of natural frequency factors for the first three modes versus ratio of G_2'/G_1 for $n = 1$.

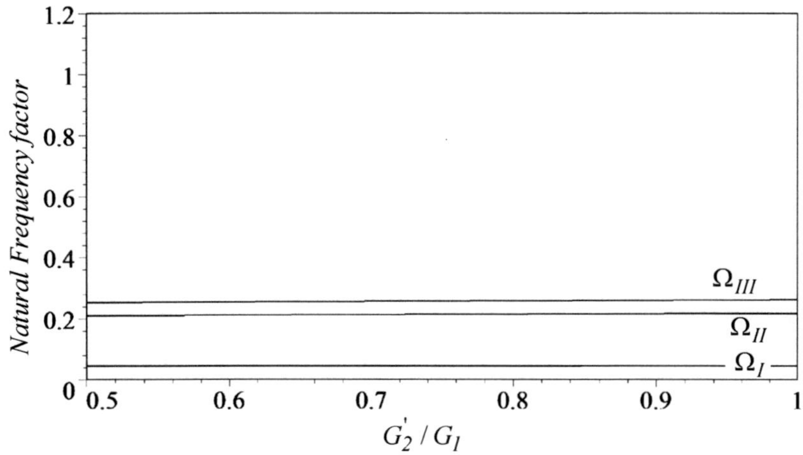

FIGURE 6.14. Variation of natural frequency factors for the first three modes versus ratio of G_2'/G_1 for $n = 2$.

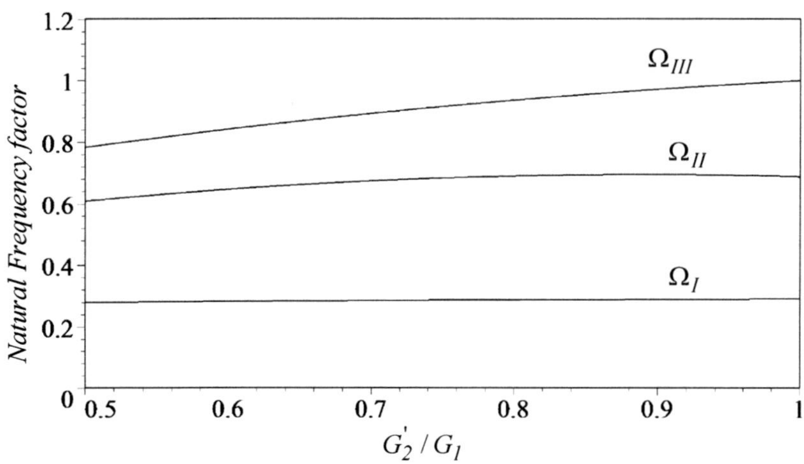

FIGURE 6.15. Variation of natural frequency factors for the first three modes versus ratio of G_2'/G_1 for $n = 3$.

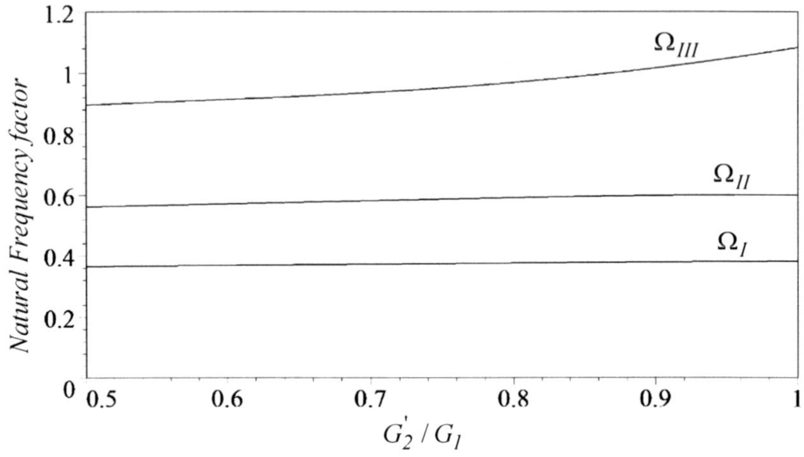

FIGURE 6.16. Variation of natural frequency factors for the first three modes versus ratio of G_2'/G_1 for $n = 4$.

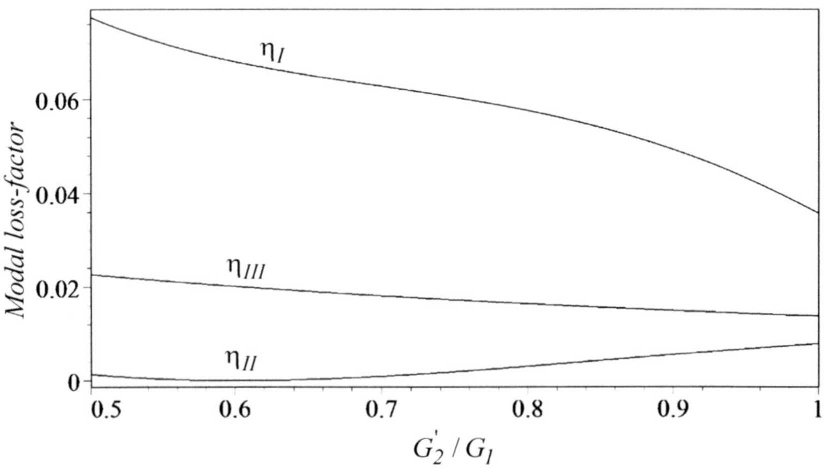

FIGURE 6.17. Variation of modal loss factors for the first three modes versus ratio of G_2'/G_1 for $n = 0$.

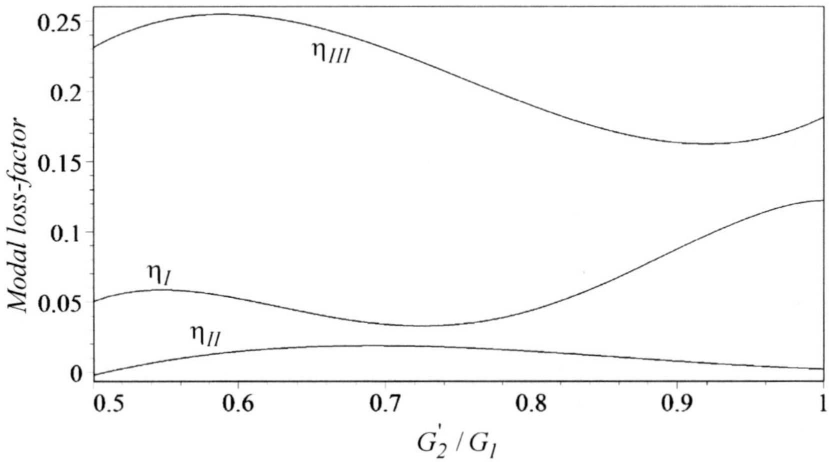

FIGURE 6.18. Variation of modal loss factors for the first three modes versus ratio of G_2'/G_1 for $n = 1$.

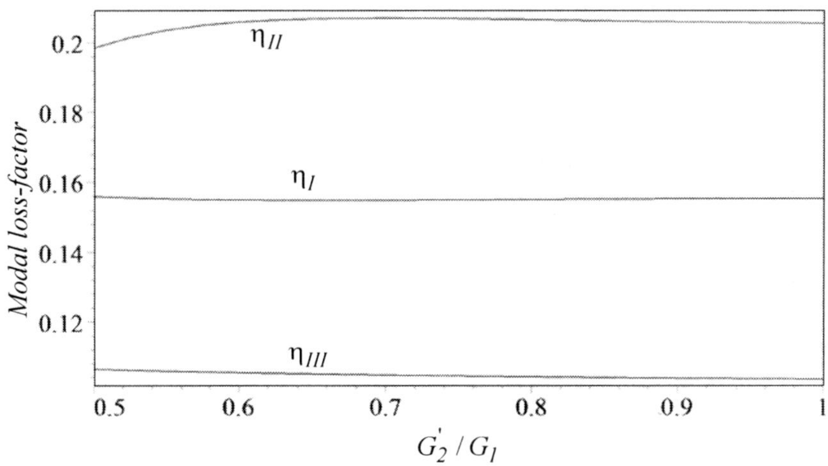

FIGURE 6.19. Variation of modal loss factors for the first three modes versus ratio of G_2'/G_1 for $n = 2$.

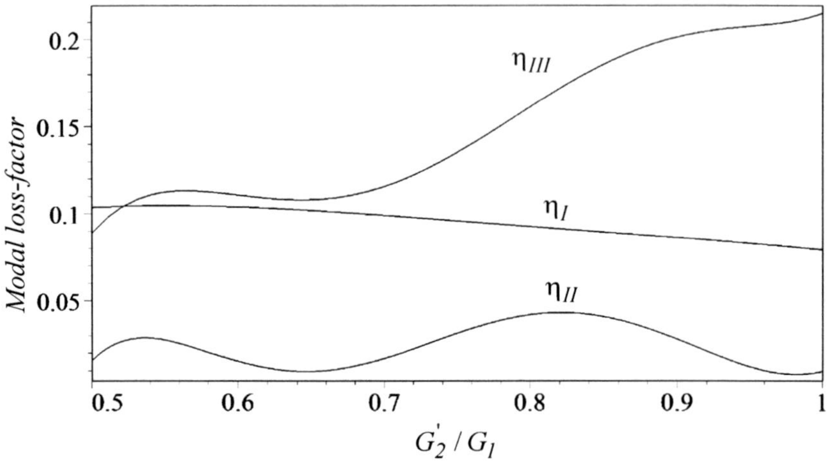

FIGURE 6.20. Variation of modal loss factors for the first three modes versus ratio of G_2'/G_1 for $n = 3$.

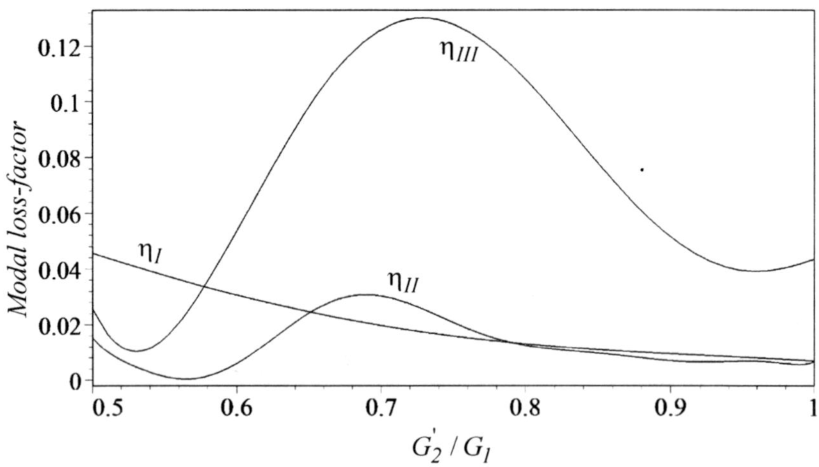

FIGURE 6.21. Variation of modal loss factors for the first three modes versus ratio of G_2'/G_1 for $n = 4$.

$r_0 = 1.0\,\text{in},$
$r_1 = 1.15\,\text{in},$
$r_2 = 1.2\,\text{in},$ and
$r_3 = 1.35924\,\text{in}.$

Thicknesses are:

$H_1 = 0.15\,\text{in},$ and
$H_3 = 0.15924\,\text{in}.$

Layer I and III were assumed to have the same properties of steel.

$G_1 = 1.1538 \times 10^7\,\text{lb/in}^2,$
$G_2 = G_2'(1 + i\,\eta_2),$
$G_2' = G_1,$ and
$\nu_1 = \nu_2 = \nu_3 = 0.3.$

For this case, it is concluded that most frequency factors changed linearly when the middle layer thickness (H_2) varied. For circumferential wave number $n = 0$, all six modal loss-factors increased linearly when the core thickness increased. For $n \geq 1$, when the thickness of the middle layer increased, the loss-factors appeared linearly either increasing or decreasing for the lower modes (mode $= 1, 2$ and 3), but for higher modes (mode $= 4, 5$ and 6), the loss-factors changed randomly.

Using the data of Tables 6.11 to 6.15, figures 6.22 to 6.26 depict the natural frequency factor Ω as a function of thickness of visco-elastic layer (in) for the first six modes with respect to the thickness of the middle layer for $n = 0$, $n = 1$, $n = 2$, $n = 3$, $n = 4$. Figures 6.27 to 6.31 show the modal loss factor η as a function of G_2'/G_1 for $n = 0$, $n = 1$, $n = 2$, $n = 3$, $n = 4$.

Table 6.11. Variation of dimensionless frequency factors and modal loss-factors for the first six modes with respect to the thickness of middle layer.

$n = 0$						
	I		II		III	
H_2 [in]	Ω	η	Ω	η	Ω	η
0.004	0.15307	0.00068	0.23105	0.00039	1.02351	0.00150
0.010	0.15333	0.00164	0.23293	0.00097	1.02369	0.00371
0.016	0.15357	0.00254	0.23481	0.00154	1.02387	0.00585
0.022	0.15281	0.00337	0.23671	0.00209	1.02405	0.00794
0.028	0.15403	0.00416	0.23860	0.00263	1.02422	0.00997
0.034	0.15423	0.00491	0.24049	0.00315	1.02439	0.01194
0.040	0.15443	0.00561	0.24239	0.00367	1.02457	0.01386
0.046	0.15462	0.00629	0.24430	0.00417	1.02473	0.01572
0.052	0.15481	0.00694	0.24620	0.00466	1.02490	0.01753
0.062	0.15509	0.00796	0.24937	0.00546	1.02517	0.02042
0.072	0.15535	0.00892	0.25254	0.00623	1.02544	0.02316
0.082	0.15561	0.00869	0.25595	0.00625	1.02590	0.02523
0.092	0.15583	0.01066	0.25888	0.00768	1.02595	0.02824
0.102	0.15604	0.01147	0.26204	0.00837	1.02620	0.03058
0.112	0.15624	0.01223	0.26520	0.00904	1.02645	0.03279
0.122	0.15642	0.01296	0.26835	0.00968	1.02670	0.03489
0.142	0.15676	0.01433	0.27463	0.01091	1.02718	0.03874
0.152	0.15692	0.01497	0.27776	0.01149	1.02741	0.04051
0.162	0.15706	0.01558	0.28087	0.01205	1.02765	0.04218
0.172	0.15720	0.01617	0.28398	0.01260	1.02788	0.04376
0.182	0.15733	0.01673	0.28707	0.01313	1.02811	0.04525
0.192	0.15745	0.01728	0.29015	0.01365	1.02835	0.04665
0.200	0.15754	0.01770	0.29262	0.01405	1.02953	0.04771

Table 6.11. Continued

$n = 0$						
	IV		*V*		*VI*	
H_2 [in]	Ω	η	Ω	η	Ω	η
0.004	1.03566	0.00078	1.86275	0.00054	2.01836	0.00000
0.010	1.03627	0.00194	1.86307	0.00135	2.01867	0.00001
0.016	1.03688	0.00306	1.86340	0.00213	2.01898	0.00002
0.022	1.03750	0.00416	1.86373	0.00289	2.01930	0.00003
0.028	1.03812	0.00523	1.86406	0.00364	2.01962	0.00006
0.034	1.03873	0.00627	1.86440	0.00438	2.01993	0.00009
0.040	1.03935	0.00729	1.86474	0.00509	2.02026	0.00014
0.046	1.03997	0.00828	1.86508	0.00580	2.02059	0.00021
0.052	1.04059	0.00925	1.86543	0.00648	2.02091	0.00029
0.062	1.04162	0.01080	1.86600	0.00761	2.02146	0.00045
0.072	1.04266	0.01229	1.86658	0.00869	2.02203	0.00066
0.082	1.04302	0.01349	1.86987	0.00876	2.02222	0.00091
0.092	1.04475	0.01508	1.86772	0.01078	2.02316	0.00120
0.102	1.04475	0.01638	1.86829	0.01178	2.02374	0.00154
0.112	1.04684	0.01763	1.86885	0.01276	2.02432	0.00192
0.122	1.04790	0.01882	1.86942	0.01371	2.02492	0.00233
0.142	1.05001	0.02105	1.87052	0.01553	2.02611	0.00326
0.152	1.05107	0.02209	1.87106	0.01641	2.02672	0.00377
0.162	1.05214	0.02309	1.87160	0.01726	2.02733	0.00431
0.172	1.05320	0.02404	1.87214	0.01809	2.02795	0.00487
0.182	1.05427	0.02495	1.87267	0.01890	2.02857	0.00545
0.192	1.05534	0.02583	1.87319	0.01969	2.02919	0.00606
0.200	1.05621	0.02650	1.87361	0.02030	2.02969	0.00655

Table 6.12. Variation of dimensionless frequency factors and modal loss-factors for the first six modes with respect to thr thickness of middle layer.

$n = 1$						
	I		*II*		*III*	
H_2 [in]	Ω	η	Ω	η	Ω	η
0.004	0.08151	0.15979	0.20183	0.18635	1.06902	0.00322
0.010	0.08127	0.15967	0.20287	0.18007	1.07328	0.00777
0.022	0.08081	0.15931	0.20495	0.17181	1.08223	0.01596
0.028	0.08058	0.15905	0.20598	0.16661	1.08676	0.01978
0.034	0.08035	0.15873	0.20700	0.16137	1.09124	0.02354
0.040	0.08012	0.15836	0.20803	0.15607	1.09567	0.02733
0.046	0.07990	0.15793	0.20904	0.15071	1.10000	0.03126
0.052	0.07968	0.15744	0.21005	0.14528	1.10425	0.03547
0.062	0.07932	0.15649	0.21173	0.13604	1.11116	0.04352
0.072	0.07897	0.15538	0.21349	0.12651	1.11826	0.05360
0.082	0.07868	0.15396	0.21545	0.11536	1.12178	0.07361
0.092	0.07831	0.15266	0.21694	0.10629	1.13701	0.07764
0.102	0.07800	0.15107	0.21868	0.09540	1.14893	0.08726
0.112	0.07771	0.14935	0.22045	0.08385	1.16145	0.09441
0.122	0.07743	0.14751	0.22225	0.07158	1.17411	0.09961
0.142	0.07692	0.14351	0.22613	0.04741	1.19920	0.10600
0.152	0.07668	0.14138	0.22825	0.03907	1.21153	0.10772
0.162	0.07646	0.13918	0.23043	0.03420	1.22370	0.10869
0.172	0.07626	0.13693	0.23260	0.03154	1.23571	0.10900
0.182	0.07606	0.13463	0.23474	0.03010	1.24775	0.10873
0.192	0.07588	0.13231	0.23685	0.02935	1.25922	0.10793
0.200	0.07575	0.13042	0.23851	0.02905	1.26843	0.10693

Table 6.12. Continued

$n = 1$						
	IV		*V*		*VI*	
H_2 [in]	Ω	η	Ω	η	Ω	η
0.004	1.35625	0.07945	1.78069	0.00197	2.08038	0.00361
0.010	1.36786	0.07636	1.79970	0.00534	2.09026	0.00732
0.022	1.39100	0.06941	1.83752	0.01447	2.11448	0.01389
0.028	1.40253	0.06548	1.85557	0.02072	2.12768	0.01717
0.034	1.41404	0.06121	1.87219	0.02824	2.14146	0.02035
0.040	1.42553	0.05648	1.88636	0.03702	2.15570	0.02328
0.046	1.43699	0.05117	1.89628	0.04717	2.17024	0.02589
0.052	1.44837	0.04461	1.88215	0.02566	2.18493	0.02814
0.062	1.46613	0.03285	1.92166	0.02001	2.20940	0.03108
0.072	1.47377	0.02581	1.95467	0.01903	2.23346	0.03312
0.082	1.49075	0.02181	1.99029	0.01838	2.25879	0.03445
0.092	1.50162	0.01877	2.01863	0.01385	2.27832	0.03525
0.102	1.51658	0.01560	2.04999	0.01280	2.29772	0.03731
0.112	1.53127	0.01226	2.08118	0.01189	2.31921	0.04217
0.122	1.54547	0.00943	2.11205	0.01123	2.34382	0.04508
0.142	1.57347	0.00723	1.95068	0.01079	2.39354	0.04450
0.152	1.58803	0.00772	1.95977	0.00967	2.41741	0.04162
0.162	1.60319	0.00895	1.96918	0.00824	2.43992	0.03722
0.172	1.62314	0.00934	1.97884	0.00756	2.46024	0.03169
0.182	1.63891	0.01076	1.98870	0.00734	2.47787	0.02612
0.192	1.65280	0.01189	1.99869	0.00707	2.49363	0.02151
0.200	1.67312	0.01295	2.00675	0.00684	2.55309	0.01712

Table 6.13. Variation of dimensionless frequency factors and modal loss-factors for the first six modes with respect to thr thickness of middle layer.

$n = 2$						
	I		*II*		*III*	
H_2 [in]	Ω	η	Ω	η	Ω	η
0.004	0.03612	0.02100	0.28192	0.21380	0.45140	0.00356
0.010	0.03588	0.02106	0.28431	0.21350	0.45732	0.00658
0.016	0.03566	0.02110	0.28670	0.21321	0.46270	0.00869
0.022	0.03544	0.02113	0.28907	0.21292	0.46786	0.01032
0.028	0.03523	0.02118	0.29143	0.21264	0.47289	0.01325
0.034	0.03503	0.02118	0.29377	0.21236	0.47785	0.01266
0.040	0.03483	0.02119	0.29611	0.21209	0.48276	0.01351
0.046	0.03465	0.02119	0.29844	0.21182	0.48764	0.01419
0.052	0.03447	0.02119	0.30076	0.21156	0.49249	0.01474
0.062	0.03419	0.02116	0.30461	0.21113	0.50055	0.01538
0.072	0.03393	0.02112	0.30842	0.21071	0.50859	0.01575
0.082	0.03355	0.02126	0.31246	0.21099	0.51735	0.01588
0.092	0.03347	0.02099	0.31596	0.21989	0.52470	0.01584
0.102	0.03326	0.02090	0.31969	0.20949	0.53278	0.01565
0.112	0.03307	0.02080	0.32339	0.20909	0.54088	0.01534
0.122	0.03289	0.02069	0.32707	0.20871	0.54902	0.01495
0.142	0.03257	0.02042	0.33433	0.20794	0.56388	0.01401
0.152	0.03236	0.02028	0.33793	0.20757	0.57361	0.01349
0.162	0.03230	0.02012	0.34149	0.20719	0.54430	0.01285
0.172	0.03218	0.01996	0.34503	0.20682	0.55004	0.01234
0.182	0.03207	0.01979	0.34854	0.20645	0.55572	0.01189
0.192	0.03197	0.01962	0.35203	0.20609	0.56134	0.01145
0.200	0.03190	0,01948	0.35480	0.20580	0.56684	0.01103

Table 6.13. Continued

$n = 2$						
	IV		V		VI	
H_2 [in]	Ω	η	Ω	η	Ω	η
0.004	0.97861	0.00376	1.17813	0.03465	1.95607	0.11676
0.010	0.97626	0.01032	1.18625	0.04732	1.96102	0.12725
0.016	0.97526	0.01796	1.19525	0.05550	1.96654	0.13506
0.022	0.97561	0.02651	1.20452	0.06151	1.97257	0.14096
0.028	0.97738	0.03559	1.21389	0.06623	1.97907	0.14538
0.034	0.98061	0.04457	1.22329	0.07008	1.98599	0.14860
0.040	0.98516	0.05266	1.23265	0.07330	1.99334	0.15077
0.046	0.99074	0.05938	1.24198	0.07606	2.00112	0.15201
0.052	0.99702	0.06468	1.25124	0.07846	2.00936	0.15239
0.062	1.00842	0.07092	1.26652	0.08188	2.02425	0.15121
0.072	1.02051	0.07475	1.28160	0.08479	2.04084	0.14795
0.082	1.03473	0.07673	1.30105	0.07956	2.07763	0.14535
0.092	1.02667	0.07956	1.35451	0.07332	2.08064	0.13698
0.102	1.02357	0.08325	1.40123	0.06923	2.10415	0.13076
0.112	1.03223	0.08756	1.44827	0.06192	2.12955	0.12525
0.122	1.03767	0.09154	1.46054	0.05753	2.15606	0.12083
0.142	1.04826	0.09645	1.48445	0.05668	2.20985	0.11504
0.152	1.05345	0.10042	1.49609	0.05622	2.23635	0.11318
0.162	1.05857	0.10368	1.50752	0.05575	2.26235	0.11166
0.172	1.06364	0.10725	1.48056	0.05923	2.28781	0.11027
0.182	1.06866	0.11143	1.44406	0.06556	2.31270	0.10884
0.192	1.07363	0.11525	1.46468	0.07181	2.33704	0.10725
0.200	1.07760	0.11956	1.477540.	07653	2.35610	0.10580

Table 6.14. Variation of dimensionless frequency factors and modal loss-factors for the first six modes with respect to thr thickness of middle layer.

$n = 3$						
	I		II		III	
H_2 [in]	Ω	η	Ω	η	Ω	η
0.004	0.38216	0.11501	0.62276	0.00063	1.07528	0.00294
0.010	0.38553	0.11527	0.62862	0.00158	1.08714	0.00514
0.016	0.38888	0.11553	0.63448	0.00256	1.09863	0.00642
0.022	0.39221	0.11579	0.64034	0.00355	1.10990	0.00730
0.028	0.39353	0.11604	0.64620	0.00457	1.12105	0.00795
0.034	0.39884	0.11629	0.65206	0.00559	1.13213	0.00843
0.040	0.40212	0.11654	0.65794	0.00663	1.14315	0.00880
0.046	0.40540	0.11679	0.66382	0.00768	1.15414	0.00905
0.052	0.40865	0.11704	0.66973	0.00872	1.16510	0.00922
0.062	0.41404	0.11744	0.67960	0.01042	1.18336	0.00928
0.072	0.41939	0.11783	0.68954	0.01205	1.20161	0.00883
0.082	0.42490	0.11790	0.70087	0.01363	1.22168	0.00772
0.092	0.42995	0.11859	0.70958	0.01498	1.20182	0.00761
0.102	0.43516	0.11896	0.71969	0.01621	1.17861	0.00781
0.112	0.44034	0.11931	0.72987	0.01728	1.14354	0.00794
0.122	0.44547	0.11966	0.74009	0.01818	1.15508	0.00811
0.142	0.45561	0.12033	0.76070	0.01945	1.17786	0.00830
0.152	0.46061	0.12065	0.77108	0.01984	1.18908	0.00836
0.162	0.46558	0.12096	0.78151	0.02007	1.20020	0.00840
0.172	0.47050	0.12126	0.79199	0.02017	1.21122	0.00843
0.182	0.47538	0.12155	0.80253	0.02015	1.22214	0.00843
0.192	0.48023	0.12184	0.81312	0.02002	1.23297	0.00842
0.200	0.48407	0.12205	0.82163	0.01985	1.24312	0.00845

Table 6.14. Continued

$n = 3$						
	IV		V		VI	
H_2 [in]	Ω	η	Ω	η	Ω	η
0.004	1.35699	0.07535	1.72049	0.00322	2.01263	0.13268
0.010	1.36830	0.07831	1.74442	0.00549	2.016380	.13356
0.016	1.37931	0.08060	1.76605	0.00709	2.02045	0.13414
0.022	1.39009	0.08229	1.78711	0.00853	2.02488	0.13433
0.028	1.40068	0.08346	1.80777	0.00999	2.02968	0.13398
0.034	1.41112	0.08415	1.82821	0.01161	2.03486	0.13296
0.040	1.42142	0.08439	1.84847	0.01350	2.04040	0.13112
0.046	1.43160	0.08421	1.86851	0.01574	2.04626	0.12830
0.052	1.44168	0.08361	1.88822	0.01836	2.05238	0.12430
0.062	1.45824	0.08163	1.92001	0.02335	2.06929	0.11428
0.072	1.47453	0.07829	1.95029	0.02858	2.07329	0.09820
0.082	1.49417	0.07516	1.98424	0.03390	2.09993	0.09780
0.092	1.50628	0.06507	2.00649	0.03874	2.15886	0.08692
0.102	1.49939	0.05881	2.05408	0.04318	2.26992	0.08354
0.112	1.50936	0.05546	2.07878	0.04138	2.34362	0.08516
0.122	1.51990	0.06247	2.12916	0.03596	2.37489	0.08823
0.142	1.54321	0.07008	2.15312	0.03184	2.43099	0.09300
0.152	1.55514	0.07255	2.17490	0.02661	2.45743	0.09414
0.162	1.56706	0.07458	2.20568	0.02114	2.48322	0.09453
0.172	1.57891	0.07636	2.23125	0.01623	2.50846	0.09425
0.182	1.59067	0.07803	2.26671	0.01245	2.53320	0.09332
0.192	1.60235	0.07970	2.30163	0.01198	2.55743	0.09177
0.200	1.61165	0.08107	2.33751	0.01056	2.58256	0.08956

Table 6.15. Variation of dimensionless frequency factors and modal loss-factors for the first six modes with respect to thr thickness of middle layer.

$n = 4$						
	I		II		III	
H_2 [in]	Ω	η	Ω	η	Ω	η
0.004	0.49651	0.00042	0.72423	0.05578	0.79863	0.00005
0.010	0.50070	0.00105	0.73124	0.05559	0.80619	0.00012
0.016	0.50487	0.00167	0.73822	0.05540	0.81373	0.00017
0.022	0.50901	0.00227	0.74516	0.05512	0.82125	0.00022
0.028	0.51313	0.00287	0.75208	0.05502	0.82875	0.00023
0.034	0.51721	0.00347	0.75897	0.05492	0.83623	0.00028
0.040	0.52127	0.00405	0.76582	0.05461	0.84369	0.00033
0.046	0.52531	0.00463	0.77265	0.05443	0.85114	0.00038
0.052	0.52932	0.00520	0.77944	0.05423	0.85857	0.00042
0.062	0.53594	0.00614	0.79070	0.05390	0.87070	0.00050
0.072	0.54249	0.00707	0.80187	0.05356	0.88323	0.00055
0.082	0.54958	0.00792	0.81327	0.05327	0.89576	0.00059
0.092	0.55539	0.00891	0.82397	0.05287	0.90778	0.00064
0.102	0.56173	0.00982	0.83489	0.05252	0.92002	0.00073
0.112	0.56801	0.01074	0.84573	0.05217	0.93224	0.00084
0.122	0.57422	0.01167	0.85649	0.05181	0.94446	0.00090
0.142	0.58645	0.01358	0.87776	0.05107	0.96889	0.00132
0.152	0.59247	0.01457	0.88826	0.05069	0.98112	0.00175
0.162	0.59842	0.01560	0.89869	0.05030	0.99336	0.00204
0.172	0.60432	0.01667	0.90904	0.04992	1.00562	0.00232
0.182	0.61015	0.01781	0.91930	0.04952	0.99320	0.00254
0.192	0.61593	0.01901	0.92947	0.04912	0.98732	0.00274
0.200	0.62051	0.02004	0.93756	0.04879	0.98231	0.00296

Table 6.15. Continued

	$n = 4$					
	IV		*V*		*VI*	
H_2 [in]	Ω	η	Ω	η	Ω	η
0.004	1.50990	0.06927	1.83845	0.00035	1.93995	0.00301
0.010	1.55997	0.06820	1.84168	0.00123	1.97150	0.01266
0.016	1.62996	0.06748	1.85902	0.00695	1.99741	0.03035
0.022	1.64790	0.06621	1.87296	0.00885	2.01177	0.04894
0.028	1.66988	0.06546	1.90895	0.01016	2.01841	0.06077
0.034	1.68009	0.06326	1.93066	0.02371	2.02339	0.06547
0.040	1.70584	0.06368	1.95061	0.03577	2.02910	0.06359
0.046	1.71791	0.06353	1.96523	0.05159	2.03987	0.05531
0.052	1.72961	0.06308	1.97317	0.07087	2.05799	0.04753
0.062	1.74821	0.06204	1.98222	0.08852	2.09216	0.04111
0.072	1.76567	0.06114	1.99115	0.11002	2.12640	0.03830
0.082	1.78017	0.06107	2.00277	0.12543	2.16238	0.03788
0.092	1.79811	0.06102	2.01108	0.14594	2.19332	0.03608
0.102	1.81342	0.06209	2.02199	0.15250	2.22591	0.03537
0.112	1.82833	0.06406	2.03341	0.15710	2.25786	0.03438
0.122	1.84658	0.06698	2.04523	0.15986	2.28919	0.03255
0.142	1.87230	0.06921	2.06972	0.16034	2.35037	0.03423
0.152	1.90155	0.07135	2.08223	0.15813	2.39966	0.03374
0.162	1.97238	0.07356	2.09481	0.15424	2.43513	0.03235
0.172	2.03521	0.07546	2.10736	0.14869	2.66451	0.03017
0.182	2.05621	0.07785	2.11977	0.14085	2.68750	0.02720
0.192	2.06023	0.07912	2.13184	0.13060	2.70916	0.02343
0.200	2.06356	0.08038	2.14103	0.12845	2.72526	0.01987

6.7 Remarks

The free and forced vibrations of thick constrained multi-layered cylinders were considered and an analytical solution based on the elasto-dynamic theory was formulated to model the system. The integrated damping treatment with a constrained visco-elastic layer was developed. The dimensionless natural frequency factors and their corresponding loss-factors were determined. The numerical results were provided for different ratios of G'_2/G_1, various material loss-factors η_2, and the range of thicknesses for the visco-elastic layer.

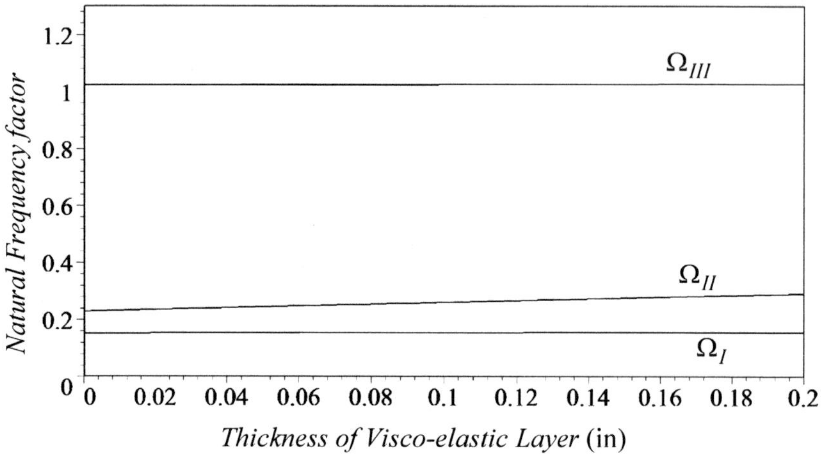

FIGURE 6.22. Variation of natural frequency factors for the first three modes versus thickness of viscoelastic layer for $n = 0$.

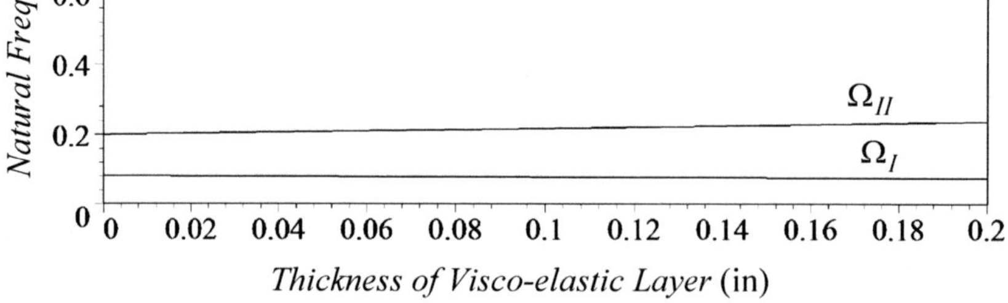

FIGURE 6.23. Variation of natural frequency factors for the first three modes versus thickness of viscoelastic layer for $n = 1$.

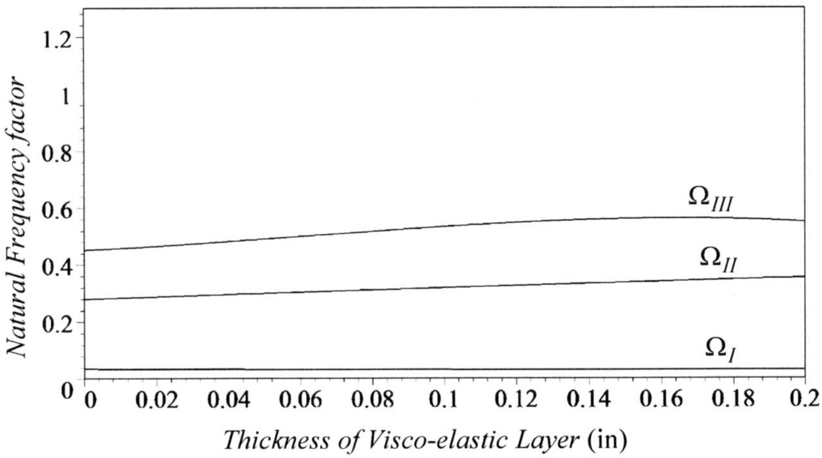

FIGURE 6.24. Variation of natural frequency factors for the first three modes versus thickness of viscoelastic layer for $n = 2$.

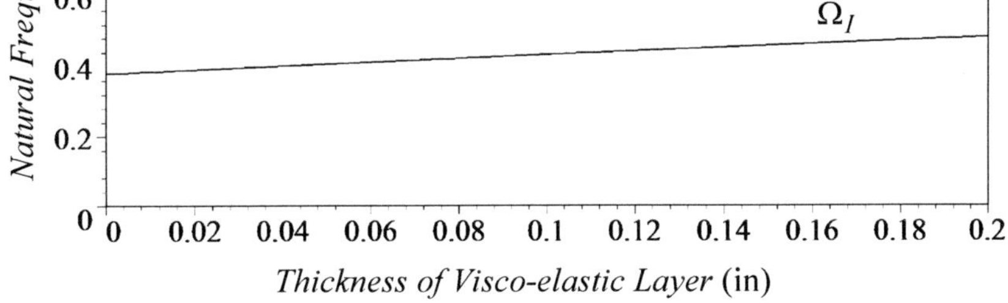

FIGURE 6.25. Variation of natural frequency factors for the first three modes versus thickness of viscoelastic layer for $n = 3$.

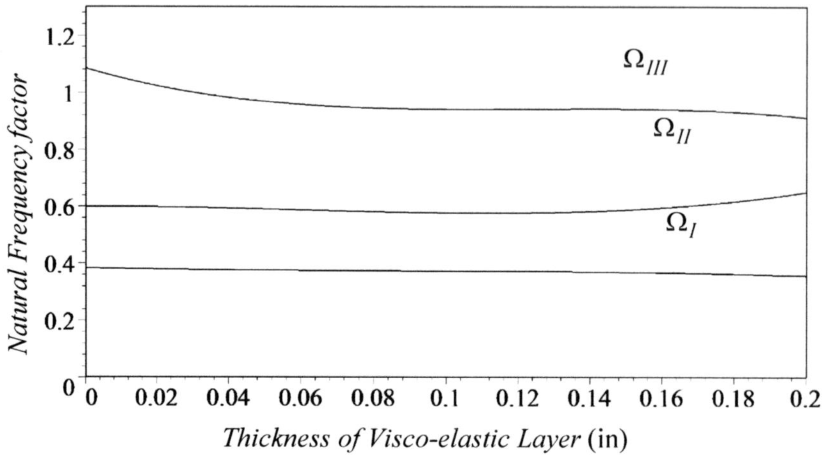

FIGURE 6.26. Variation of natural frequency factors for the first three modes versus thickness of viscoelastic layer for $n = 4$.

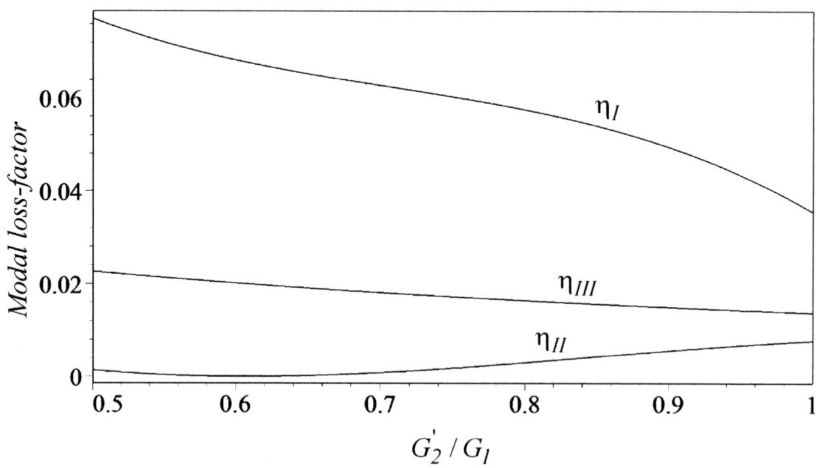

FIGURE 6.27. Variation of modal loss factors for the first three modes versus ratio of G_2'/G_1 for $n = 0$.

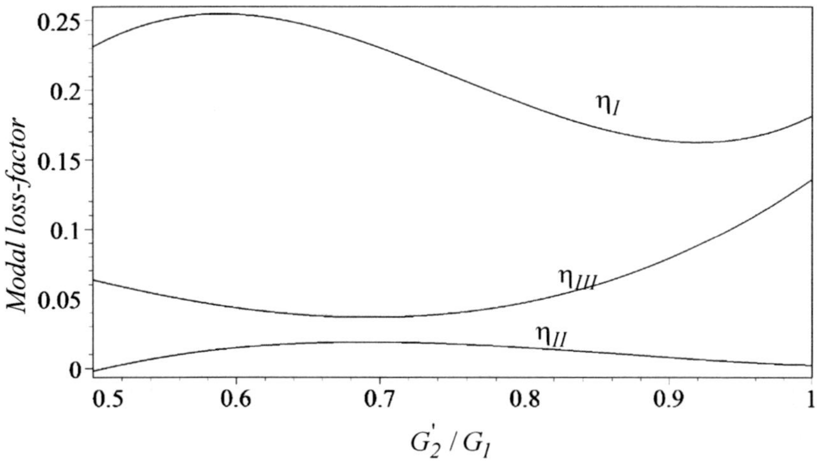

FIGURE 6.28. Variation of modal loss factors for the first three modes versus ratio of G_2'/G_1 for $n = 1$.

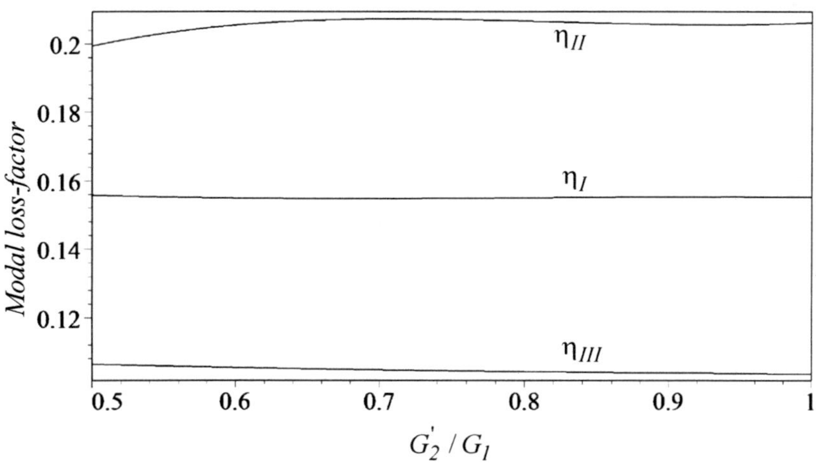

FIGURE 6.29. Variation of modal loss factors for the first three modes versus ratio of G_2'/G_1 for $n = 2$.

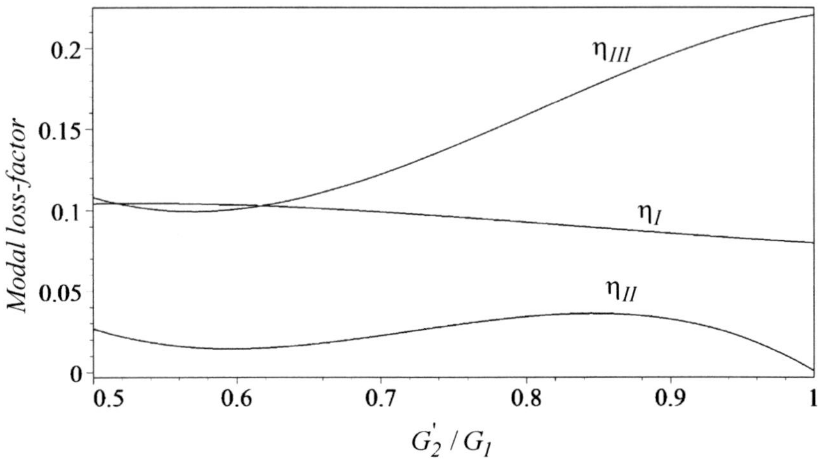

FIGURE 6.30. Variation of modal loss factors for the first three modes versus ratio of G_2'/G_1 for $n = 3$.

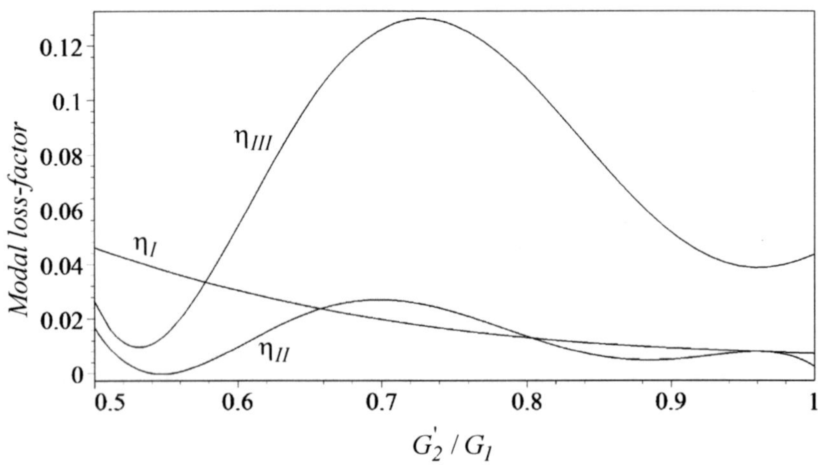

FIGURE 6.31. Variation of modal loss factors for the first three modes versus ratio of G_2'/G_1 for $n = 4$.

In the case of various material damping for the middle layer, the frequency factors were relatively unchanged while the material loss-factor increased for most modes. For axisymmetric modes, the modal loss-factors appeared to increase linearly with the increasing material loss-factor of the middle layer. When reducing the shear modulus for the middle layer, frequency factors, for $n = 0$, decreased linearly. For $n = 2$, the changes for natural frequency factors are relatively insignificant. The modal loss-factor, for all circumferential waves, appeared to vary when the shear modulus of the middle layer decreased. It should be noted that the modal loss-factors in general are relatively higher for $n = 2, 3$, and 4, when comparing with cases for $n = 2$ and $n = 1$.

The study also included the effects of variation in thickness of the constrained layer on natural frequency factors and modal loss-factors for sandwiched cylinders. It was shown that most frequency factors changed linearly when the middle layer thickness varied. For the circumferential wave number $n = 0$, all six modes of the modal loss-factors varied linearly, either increasing or decreasing, for the lower modes (mode $= 1, 2$ and 3), and for higher modes (mode $= 4, 5$ and 6), the loss-factors changed non-linearly.

6.8 Key Symbols

E	Young's modulus
E'	real parts of E
G, μ	shear modulus, modulus of rigidity
G'	real parts of G
H	thickness of the cylinder
L	axial half wavelength
n	wave number
$n = 0$	breathing mode
$n = 1$	rigid body mode
$n > 1$	lobar modes
r	radial direction
R	mean radius of the cylinder
v_2	velocity of propagation of the distortional wave
z	axial direction
$\zeta = \pi/L$	axial wave number
η	modal loss-factor
η_2	core loss-factor
θ	transverse direction
λ	Lame's elastic constant
μ, G	shear modulus, modulus of rigidity
ν	Poisson's ratio
ρ	density
σ	normal stress
τ	shear stress
ω	excitation frequency
ω_s	the lowest simple thickness shear frequency
$\Omega = \omega/\omega_s$	frequency factor
$\Im\omega$	imaginary part of ω
$\Re\omega$	real part of ω

7

Vibrations of Partial Cylindrical Panels

Free and forced vibrations of thick visco-elastic cylindrical panels are investigated. The medium in this case is assumed to be homogeneous, isotropic, linearly visco-elastic, thick cylindrical panel of infinite extent. In development of an analytical solution, two dimensional elasto-dynamic theory is employed and linear visco-elastic material is allowed by assuming complex elastic moduli. It should be noted that the panel ends condition does not allow for the displacement along the transverse direction, however, radial displacements are permitted. The surface impedance of the panel is formulated and modal characteristics of the panel are determined. This is achieved by utilizing numerical values of Bessel functions for complex argument and fractional order.

7.1 Introduction

An important step in the study of cylindrical panels is the evaluation of the modal displacement functions and their corresponding resonant frequencies for infinitely long panels. Modal information plays a key role in the design of cylindrical panels subjected to dynamic excitations. Most studies on the dynamic response of cylindrical panels have been restricted to the application of the thin shell theories. In this chapter, the equation of motion for a cylindrical shell is formed by considering two-dimensional plane strain nature of the problem and assuming that the displacement and stresses are not changing along the axis of the cylindrical panel.

To control the large displacement and stresses associated with resonance, damping treatment is considered by introducing complex elastic moduli. The visco-elastic panel dissipates much of its energy in the form of heat during shear deformation when subjected to harmonic vibrations. It should be noted that the panel end conditions do not allow the displacements along the transverse direction; however, the radial displacements are not constrained. The modal displacements and stresses at any point in the panel can be formulated in terms of the modal radial and shear stresses at the inner and the outer boundaries. Consequently for a free vibration,

H.R. Hamidzadeh, R.N. Jazar, *Vibrations of Thick Cylindrical Structures*
DOI 10.1007/978-0-387-75591-5_7,

exciting boundary stresses can be assumed to be zero and non-dimensional natural frequencies and their respective modal displacements and stresses will be determined. In the case of having linear visco-elastic material for the panel, the modal loss factors can also be computed for the breathing and flexural modes, and their respective thickness modes for different material damping factors. The cylindrical panel with different panel angles can also be modeled using finite element method.

7.2 Objectives of the Present Chapter

The general two-dimensional plane strain model is used to investigate the harmonic wave propagation in a visco-elastic cylindrical panel. The cylindrical panel shown in figure 7.1, is assumed to be homogeneous, isotropic, linearly visco-elastic and infinitely long.

The employed mathematical model is based on the reported work of Hamidzadeh (1997). He extended his research work on vibration analysis of thick cylindrical structures to develop the two-dimensional solution for the visco-elastic cylindrical panels. In this chapter, a solution is developed to determine the modal stresses and displacements for any point in the medium for panels with any angle.

The mathematical model is presented for the governing equations of motion. The analytical solution to the harmonic response of a long visco-elastic cylindrical panel subjected to exciting boundary stresses is investigated and the displacements and stresses for each mode are evaluated. The chapter also presents computed natural frequency factors for several panel angles for breathing and flexural modes, as well as, their corresponding thickness modes for a wide range of thickness to radius ratios. Also, the computation is conducted to study the effect of structural damping on the natural frequencies and the modal loss factors for panels. In addition to the results from the analytical modeling, results from the the finite element model are also presented for the non-dimensional natural frequencies.

Figure 7.1 depicts the geometry of the panel, where a and b indicate the inner and outer radius of the cylindrical panel respectively. r is the radius of the any point on the panel and ψ is the panel angle. u_r and u_θ are the radial and transverse displacements of the panel. H is defined as the thickness of the panel.

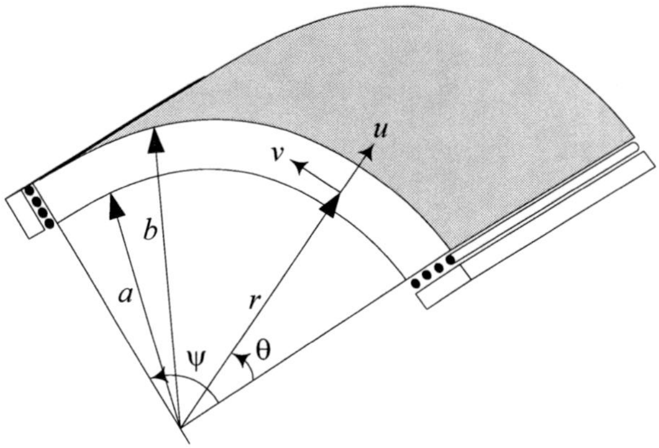

FIGURE 7.1. Coordinates and displacements of a typical thick visco-elastic cylindrical panel.

7.3 Harmonic Vibrations of Cylindrical Panels

This section presents an analytical solution to the harmonic response of an infinitely long elastic cylindrical panel with internal damping subjected to flexural forced vibration. The cylinder is assumed to be homogeneous and isotropic, linearly viscoelastic which is subjected to inner and outer boundary stresses. The equations of motion are combined in such a way that four Bessel differential equations are obtained. These equations are solved to determine the displacements. Once the displacements are known, the stress components at any point in the cylindrical panel are obtained by considering the panel to be infinitely long subjected to harmonic modal excitations. Then, by satisfying the boundary conditions, the two stresses and two displacements at any point in the cylindrical panel can be computed for any given exciting frequency. The resonant frequency for the isotropic panel can be estimated by observing the values of excited frequency when maximum amplitudes for stresses and displacements occur. The procedure to solving the equation for resonant frequencies is done in the following pages.

7.3.1 Modified Equations of Motion

Plane strain is a more general type of strain which includes simple extension and shearing as particular cases, and is obtained by assuming that one of the principal extensions is zero. If the corresponding principal axis is the axis of z, then the shearing strains $\varepsilon_{\theta z}$ and $\varepsilon_{\theta z}$ vanish and u_z .satisfies the following condition:

$$\frac{\partial u_z}{\partial z} = 0 \qquad u_r,\ u_\theta,\ and\ u_z\ independent\ of\ z \tag{7.1}$$

while u_r and u_θ are functions of r and θ only. Using the above conditions two of the three elastic rotations become zero.

$$\omega_r = \omega_\theta = 0 \tag{7.2}$$

Then governing equations of motion for the infinitely long cylindrical structure based on the theory of elasto-dynamics were presented by equations (2.54) to (2.56). Considering the plane strain case and assuming that the displacements and stresses are not varying along the axial direction, then the governing equations can be reduced to the following equations:

$$(\lambda + 2\mu)\, r\frac{\partial \varepsilon}{\partial r} - 2\mu\frac{\partial \omega_z}{\partial \theta} = -\rho p^2 r u_r \tag{7.3}$$

$$(\lambda + 2\mu)\, \frac{1}{r}\frac{\partial \varepsilon}{\partial \theta} - 2\mu\frac{\partial \omega_r}{\partial z} = -\rho p^2 r u_\theta \tag{7.4}$$

where ε and ω_z are volumetric strain and elastic rotation about the z axis and are presented by the following equations:

$$\varepsilon = \frac{\partial u}{\partial r} + \frac{1}{r}\frac{\partial u_\theta}{\partial \theta} + \frac{u_r}{r} \tag{7.5}$$

$$\omega_z = \frac{1}{2}\left(\frac{\partial u_\theta}{\partial r} + \frac{u_\theta}{r} - \frac{1}{r}\frac{\partial u_r}{\partial \theta}\right) \tag{7.6}$$

Differentiating the two governing equations of motions (7.3) and (7.4) with respect to r and θ and combining them with (7.5) and (7.6) yield:

$$(\lambda + 2\mu)\left(r\frac{\partial^2 \varepsilon}{\partial r^2} + \frac{\partial \varepsilon}{\partial r} + \frac{1}{r}\frac{\partial^2 \varepsilon}{\partial \theta^2}\right) = -\rho p^2 \varepsilon r \tag{7.7}$$

$$\mu\left(\frac{1}{r}\frac{\partial^2 \omega_z}{\partial \theta^2} + \frac{\partial \omega_z}{\partial r} + r\frac{\partial^2 \omega_z}{\partial r^2}\right) = -\rho p^2 r \omega_z \tag{7.8}$$

Two parameters α and β are introduced for simplification and are defined as:

$$\beta^2 = \frac{\rho p^2}{\lambda + 2\mu} \tag{7.9}$$

$$\alpha^2 = \frac{\rho p^2}{\mu} \tag{7.10}$$

Substituting the above parameters in equation (7.7) and (7.8), they can be simplified as:

$$r\frac{\partial^2 \varepsilon}{\partial r^2} + r\frac{\partial \varepsilon}{\partial r} + r^2\beta^2\varepsilon = -\frac{\partial^2 \varepsilon}{\partial \theta^2} \tag{7.11}$$

$$r\frac{\partial^2 \omega_z}{\partial r^2} + r\frac{\partial \omega_z}{\partial r} + r^2\alpha^2\omega_z = -\frac{\partial^2 \omega_z}{\partial \theta^2} \tag{7.12}$$

These are the modified equations of motion in terms of two elastic functions, namely the volumetric strain ε and the elastic rotation ω_z. Now simplify the parameters to get the velocity of dilatational and distortional waves. From equation

$$\frac{p^2}{\beta^2} = \frac{\lambda + 2\mu}{\rho} \tag{7.13}$$

v_1 is defined as the velocity of propagation of a dilatational wave and is given by

$$v_1 = \frac{p}{\beta}\sqrt{\frac{\lambda + 2\mu}{\rho}}. \tag{7.14}$$

Similarly,

$$\frac{p^2}{\alpha^2} = \frac{\mu}{\rho} \tag{7.15}$$

v_2 is defined as the velocity of distortional waves.

$$v_2 = \frac{p}{\alpha}\sqrt{\frac{\mu}{\rho}} \tag{7.16}$$

7.3.2 Solution to the Modified Equation of Motions

By assuming the following harmonic solution for ε and ω_z,

$$\varepsilon = \bar{\varepsilon}(r)\cos(n\theta)\, e^{i\omega t} \tag{7.17}$$

$$\omega_z = \bar{\omega}_z(r)\sin(n\theta)\, e^{i\omega t} \tag{7.18}$$

the equations (7.11) and (7.12) can be treated as separate independent Bessel differential equations for $\bar{\varepsilon}$ and $\bar{\omega}_z$. The solution to these equations is readily obtained from the theory of Bessel functions. In the above equations the notations ε_r and ω_z are functions of r only and

$\omega =$ circular frequency
$n =$ number of circumferential wave

Substituting equation (7.17) in (7.11) and (7.18) in (7.12) yield:

$$r^2\frac{\partial^2\bar{\varepsilon}}{\partial r^2} + r\frac{\partial\bar{\varepsilon}}{\partial r} + \left(r^2\beta^2 - n^2\right)\bar{\varepsilon} = 0 \tag{7.19}$$

$$r^2\frac{\partial^2\bar{\omega}_z}{\partial r^2} + r\frac{\partial\bar{\omega}_z}{\partial r} + \left(r^2\alpha^2 - n^2\right)\bar{\omega}_z = 0 \tag{7.20}$$

The above two equations are recognized as Bessel equations whose solutions are given by:

$$\bar{\varepsilon}(r) = A_n J_n(\beta r) + B_n Y_n(\beta r) \tag{7.21}$$

$$\bar{\omega}_z(r) = C_n J_n(\alpha r) + D_n Y_n(\alpha r) \tag{7.22}$$

In the above equations J_n and Y_n represent the Bessel functions of the first and second kind or the modified Bessel function of order n of the first and second kind. The choice of Bessel function depends on whether the argument is real or imaginary. If the argument is positive, J_n and Y_n are used; on the other hand, if the argument is negative, then I_n and K_n are used. The proper selection of Bessel functions for this case is very important.

By using equations (7.21) and (7.22) the above equations can be written by:

$$\varepsilon = [A_n J_n(\beta r) + B_n Y_n(\beta r)] \cos(n\theta)\, e^{i\omega t} \tag{7.23}$$

$$\omega_z = [C_n J_n(\alpha r) + D_n Y_n(\alpha r)] \sin(n\theta)\, e^{i\omega t} \tag{7.24}$$

7.3.3 Modal Displacements

In this section the relations among stresses, displacements and elastic rotations developed in chapter 2 are used along with the solution of the governing equations of motion. The equations for displacements are derived in terms of Bessel functions. From equations (7.3) and (7.4), the harmonic displacements can be written by:

$$u_r = -\frac{1}{\beta^2}\frac{\partial \varepsilon}{\partial r} + \frac{2}{r}\frac{1}{\alpha^2}\frac{\partial \omega_z}{\partial \theta} \tag{7.25}$$

$$u_\theta = -\frac{1}{r\beta^2}\frac{\partial \varepsilon}{\partial \theta} - \frac{1}{\alpha^2}\frac{\partial \omega_z}{\partial r} \tag{7.26}$$

Using equation (7.23) and (7.24) along with their appropriate derivatives equations (7.25) and (7.26) can be rewritten by the following equations:

$$u_r = \left(-\frac{1}{\beta}\left(A_n J_n'(\beta r) + B_n Y_n'(\beta r)\right) + \frac{2n}{r\alpha^2}\left(C_n J_n(\alpha r) + D_n Y_n(\alpha r)\right)\right) \cos(n\theta)\, e^{i\omega t} \tag{7.27}$$

$$u_\theta = \left(-\frac{2}{\alpha}\left(C_n J_n'(\alpha r) + D_n Y_n'(\alpha r)\right) + \frac{n}{r\beta^2}\left(A_n J_n(\beta r) + B_n Y_n(\beta r)\right)\right) \sin(n\theta)\, e^{i\omega t} \tag{7.28}$$

7.3.4 Modal Stresses

As discussed in chapter two, the amplitudes of stresses on the plane normal to the radial axis in the elastic medium, in terms of volumetric strain ε and elastic rotation ω_z are:

$$\sigma_{rr} = \lambda\varepsilon + 2\mu\frac{\partial u}{\partial r} \tag{7.29}$$

$$\tau_{r\theta} = 2\mu\left(\frac{\partial u_\theta}{\partial r} - \omega z\right) \tag{7.30}$$

Substituting from the equations (7.23) and (7.24), (7.29) and (7.30), the component of stresses will be

$$\sigma_{rr_n} = (A_n E_n(r) + B_n F_n(r) + C_n G_n(r) + D_n H_n(r)) \cos(n\theta) \tag{7.31}$$

$$\tau_{r\theta_n} = (A_n E_n^*(r) + B_n F_n^*(r) + C_n G_n^*(r) + D_n H_n^*(r)) \sin(n\theta) \tag{7.32}$$

In the above equations the coefficients E_n, F_n, G_n, H_n, E_n^*, F_n^*, G_n^*, and H_n^* are functions of Bessel functions, where:

$$E_n(r) = \lambda J_n(r\beta) - 2\mu J_n''(r\beta) \tag{7.33}$$

$$F_n(r) = \lambda Y_n(r\beta) - 2\mu Y_n(r\beta) \tag{7.34}$$

$$G_n(r) = 4\mu n\frac{\mu}{r\alpha} J_n'(r\alpha) - 4\mu n\frac{1}{r^2\alpha^2} J_n(r\alpha) \tag{7.35}$$

$$H_n(r) = 4\mu n\frac{\mu}{r} Y_n'(r\alpha) - 4\mu n\frac{1}{r^2\alpha^2} Y_n(r\alpha) \tag{7.36}$$

$$E_n^*(r) = 2\mu n\frac{1}{r\beta} J_n'(r\beta) - 2\mu n\frac{1}{r^2\beta^2} J_n(r\beta) \tag{7.37}$$

$$F_n^*(r) = 2\mu n\frac{1}{r\beta} Y_n'(r\beta) - 2\mu n\frac{1}{r^2\beta^2} Y_n(r\beta) \tag{7.38}$$

$$G_n^*(r) = -4\mu J_n''(r\alpha) - 2\mu J_n(r\alpha) \tag{7.39}$$

$$H_n^*(r) = -4\mu Y_n''(r\alpha) - 2\mu Y_n(r\alpha) \tag{7.40}$$

Functions $J_n'(r)$, $J_n''(r)$, $Y_n'(r)$, and $Y_n''(r)$ are the first and second derivatives of $J_n(r)$ and $Y_n(r)$ with respect to r.

Considering the boundary stresses in inner and outer surfaces, these stresses can be presented as:

$$\begin{bmatrix} E_n(a) & F_n(a) & G_n(a) & H_n(a) \\ E_n^*(a) & F_n^*(a) & G_n^*(a) & H_n^*(a) \\ E_n(b) & F_n(b) & G_n(b) & H_n(b) \\ E_n^*(b) & F_n^*(b) & G_n^*(b) & H_n^*(b) \end{bmatrix} \begin{Bmatrix} A_n \\ B_n \\ C_n \\ D_n \end{Bmatrix} = \begin{Bmatrix} \sigma_{rr_n}(a) \\ \tau_{r\theta_n}(a) \\ \sigma_{rr_n}(b) \\ \tau_{r\theta_n}(b) \end{Bmatrix} \tag{7.41}$$

or

$$T_n X_n = B_n \tag{7.42}$$

where B_n is a vector containing radial and shear stresses on the inner and outer surface of the medium, X_n is a vector containing arbitrary constants; and T_n is a square matrix containing the coefficients in terms of Bessel functions.

7.4 Modal Displacement and Stresses

To determine the displacements and stress components at any given point in the cylinder, the above equation can be arranged in the following matrix equation,

$$R_n(r) = S_n(r) X_n \tag{7.43}$$

where $R_n(r)$ is a vector containing the components of radial and transverse displacement and stress, and $S_n(r)$ is a coefficient square matrix.

$$R_n(r) = \begin{Bmatrix} \dfrac{u_{r_n}(r,\theta)}{\cos(n\theta)} \\ \dfrac{u_{\theta_n}(r,\theta)}{\sin(n\theta)} \\ \dfrac{\sigma_{rr_n}(r,\theta)}{\cos(n\theta)} \\ \dfrac{\tau_{r\theta_n}(r,\theta)}{\sin(n\theta)} \end{Bmatrix} \tag{7.44}$$

$$S_n(r) = \begin{bmatrix} -\dfrac{1}{\beta}J_n'(r\beta) & -\dfrac{1}{\beta}Y_n'(r\beta) & \dfrac{2n}{r\alpha^2}J_n(r\alpha) & \dfrac{2n}{r\alpha^2}Y_n(r\alpha) \\ \dfrac{n}{r\beta^2}J_n(r\beta) & \dfrac{n}{r\beta^2}Y_n(r\beta) & -\dfrac{2}{\alpha}J_n'(r\mu) & -2\dfrac{2}{\alpha}Y_n'(r\mu) \\ E_n(r) & F_n(r) & G_n(r) & H_n(r) \\ E_n^*(r) & F_n^*(r) & G_n^*(r) & H_n^*(r) \end{bmatrix} \tag{7.45}$$

$$X_n = \begin{Bmatrix} A_n \\ B_n \\ C_n \\ D_n \end{Bmatrix} \tag{7.46}$$

Arranging the equations (7.42)-(7.43) and eliminating X_n, yields:

$$D_n(r)\, R_n(r) = \sigma_n \tag{7.47}$$

where,

$$D_n(r) = T_n S_n^{-1}(r) \tag{7.48}$$

For a given lobar boundary stress component, the above equation can provide the displacements and stress components at any point in the medium.

7.5 Modal Analysis

There are a number of structures, from turbine blades to suspension bridges for which structural integrity is a real concern, and for which a thorough and precise knowledge of dynamic characteristics is essential. There are even wider sets of components or assemblies for which vibration levels encountered in service can be anticipated and brought under satisfactory control. The major application of modal analysis is to identify the vibration modes which are desired to validate the theoretical model prior to its use for predicting response levels to complex excitations. It is generally felt that the corroboration of the major modes of vibration can provide reassurance of the basic validity of the model which may then be put to further use. From the theoretical modes, the damping in a mode is predicted and can be compared with the modal damping in the system.

The first five dimensionless resonant frequency factors are computed by determining the frequency factor associated with the maximum absolute value of the modal displacement. This has been done independently for the first five modes of vibration. Different parameters of the panel geometry were varied and the frequency factors were computed. The dominance of a particular mode depends upon the type of excitation. The mode of the vibration of panels is classically defined as flexural mode, where there is a transverse displacement. When there is no transverse displacement, it is known as breathing mode. Figures 7.2 to 7.4 show the mode shapes for different n values.

Mathematically, for each mode of vibration, there are an infinite number of solutions. This is due to the nature of Bessel differential equations and

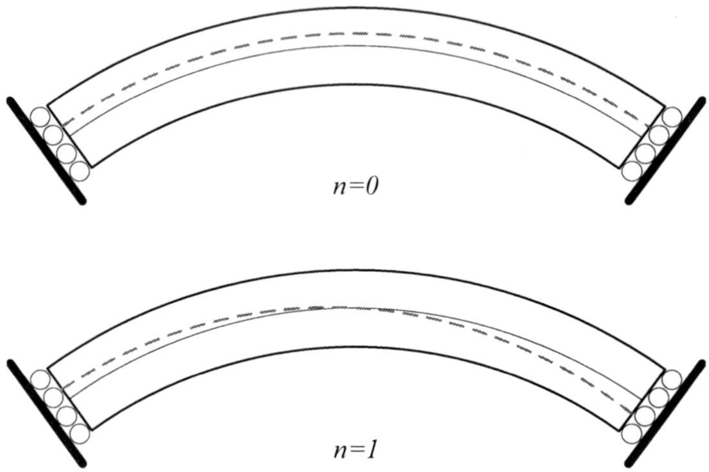

FIGURE 7.2. Modal displacement patterns for thick panel for $n = 0$ and $n = 1$.

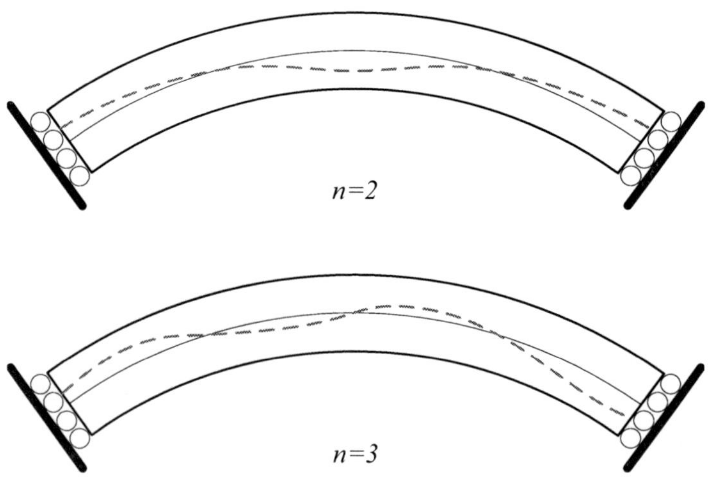

FIGURE 7.3. Modal displacement patterns for a thick panel when $n = 2$ and $n = 3$.

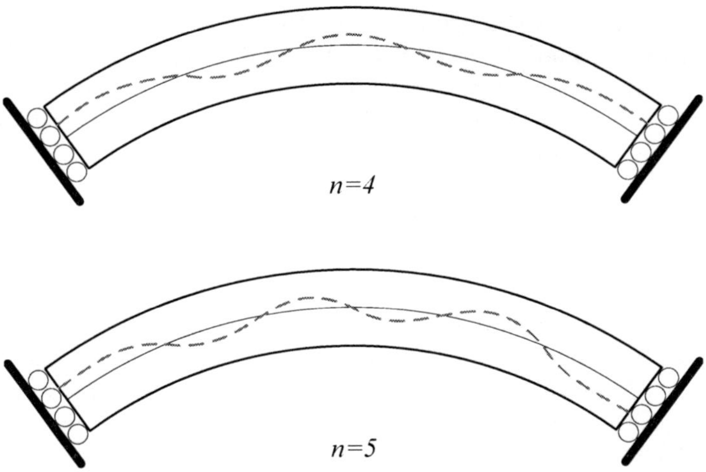

FIGURE 7.4. Modal displacement patterns for thick panel for $n = 4$ and $n = 5$.

their solutions. The thickness modes are the solution for each mode of vibration. Whenever there are displacements of a continuous system caused by excitations of the resonant frequency, nodal lines appear. These nodal lines describe the thickness modes.

7.6 Computed Resonant Frequencies

The following values and range of parameters are used in the computation:

Poisson's ratio $\nu = 0.25$
Ratio of thickness to mean radius to $H/R = 0.3$ to 1.9
Circumferential wave numbers $n = 1, 2, 3, 4$
Density $\rho = 0.000725\,\mathrm{lb\,s^2/\,in^2}$
Young modulus $E = 30000000\,\mathrm{lb/\,in^2}$
Damping loss-factor $\eta = 0.0, 0.05, 0.1$
Panel angle ranging from 30 deg to 180 deg

To identify the approximate dimensionless resonant frequency (frequency factor $= \beta H$, where H is the thickness of the panel), a frequency sweep is made for a set of geometrical and material parameters. The frequency factors are computed by determining the factors associated with the maximum value of modal displacements. To determine the modal parameters, a small frequency factor is considered to assure accurate estimation of resonant frequencies. In the analysis, modal radial boundary stresses with amplitude of 1000psi are applied on the inner boundary and the amplitudes of other boundary stresses are assumed to be zero. The modes are computed for different panel angles.

The dominance of a particular mode depends upon the type of excitation. The modes of vibration for a panel are classically defined as flexural modes, where there are tangential displacements. When there is not a tangential displacement, then it is known as breathing mode. Mathematically, for each mode of vibration, there is an infinite number of solution. This is due to the nature of Bessel differential equations and their solutions. The thickness modes are the solution for each mode of vibration. Whenever there are displacements of a continuous system, caused by excitations of the resonant frequency, nodal lines appear. These nodal lines describe the thickness modes. Discussion of the results follows the respective set of tables.

The resonant frequency factor for five modes for panel angles, varying from 30 deg to 180 deg, are presented in Tables 7.1 through 7.4 for $n = 1, 2, 3$ and 4, and $H/R = 0.3$. Figures 7.4 through 7.7 show how these factors vary for different panel angles. Furthermore, the frequency response are provided for angles of 30 deg through 180 deg, for panel with mode 2 and $H/R = 0.3$.

The resonant frequency factors presented in this chapter are in similar with those of Armenakas et. al. (1969). The results for $n = 0$ is independent of the angle of the panel and is as those of thick circular cylinders which are tabulated by Armenakas et. al. (1969). The resonant frequency factors increases with the increasing angle of panel. The modes of the panel are classically defined as flexural modes when $m > 0$, and axially symmetric extensional modes when $m = 0$. Figure 7.5 shows the flexural modes and their associated thickness modes.

The fundamental frequency factors for five modes for panel angle changing from 40 to 180 degree with increments of 20 deg computed are tabulated in Tables 7.5 through 7.13. Figures 7.5 through 7.16 shows the graphical results. The computations were done for different H/R ratios changing form 0.1 to 1.0 with an increment of 0.1. The fundamental frequency factors increased with the increase in thickness.

Table 7.1. Variation of the first five resonant frequency factors with respect to the panel angles for a circumferential number $n = 1$, $H/R = 0.3$.

ψ	Resonant Frequency Factors (βH)				
30 deg	0.48777	1.43670	2.44850	2.97860	4.14000
40 deg	0.29370	1.10980	2.21500	3.01280	3.96050
50 deg	0.19017	0.89543	2.09180	3.04480	3.86400
60 deg	0.12974	0.76006	2.01870	3.06890	3.80450
70 deg	0.09183	0.66401	1.97340	3.08610	3.76590
80 deg	0.06596	0.59339	1.94300	3.09890	3.73920
90 deg	0.04815	0.53975	1.92160	3.10860	3.71980
100 deg	0.03460	0.49757	1.90607	3.11600	3.70490
110 deg	0.02605	0.46389	1.89445	3.12175	3.69490
120 deg	0.01870	0.43652	1.88551	3.12635	3.68740
130 deg	0.13510	0.41397	1.87851	3.13003	3.67990
140 deg	0.00974	0.39516	1.8729	3.13551	3.67490
150 deg	0.00670	0.37931	1.86839	3.13761	3.66990
160 deg	0.00390	0.36582	1.86467	3.13761	3.66670
170 deg	0.00220	0.35427	1.86157	3.13930	3.66380
180 deg	0.00000	0.34430	1.85893	3.14701	3.66150

Table 7.2. Variation of the first five resonant frequency factors with respect to the panel angles for a circumferential number $n = 2$, $H/R = 0.3$.

	Resonant Frequency Factors (βH)				
ψ	1	2	3	4	5
30 deg	1.36810	2.52870	3.22540	3.63270	4.97230
40 deg	0.91823	2.06090	2.92820	3.07680	4.54430
50 deg	0.65538	1.69730	2.6529	2.97660	4.29640
60 deg	0.48786	1.43520	2.24489	2.97860	4.14110
70 deg	0.37409	1.24490	2.31080	2.99500	4.03660
80 deg	0.29372	1.09980	2.21400	3.01350	3.96210
90 deg	0.23470	0.98660	2.14370	3.03040	3.90600
100 deg	0.19069	0.89528	2.09080	3.04490	3.86380
110 deg	0.15663	0.82147	2.05040	3.05760	3.83090
120 deg	0.12979	0.76010	2.01900	3.06830	3.80470
130 deg	0.10875	0.70826	1.98480	3.07830	3.78060
140 deg	0.91795	0.66437	1.97360	3.08630	3.76660
150 deg	0.07733	0.62646	1.95660	3.09280	3.75160
160 deg	0.06625	0.59354	1.94320	3.10230	3.73920
170 deg	0.05638	0.56511	1.93120	3.10420	3.72920
180 deg	0.04785	0.53967	1.92170	3.10880	3.72020

Table 7.3. Variation of the first five resonant frequency factors with respect to the panel angles for a circumferential number $n = 3$, $H/R = 0.3$.

	Resonant Frequency Factors (βH)				
ψ	1	2	3	4	5
30 deg	2.25820	3.09470	4.20740	4.84595	5.79520
40 deg	1.59630	2.68770	3.44510	3.94181	5.18300
50 deg	1.18730	2.36710	3.08870	3.39250	4.80140
60 deg	0.91846	2.06900	2.92770	3.07661	4.54330
70 deg	0.72970	1.80500	2.74110	2.98640	4.36690
80 deg	0.59160	1.60090	2.57500	2.97430	4.23870
90 deg	0.48780	1.40210	2.44890	2.97860	4.14180
100 deg	0.40750	1.30280	2.35110	2.98950	4.06810
110 deg	0.34420	1.19230	2.27470	3.00120	4.00890
120 deg	0.29371	1.10020	2.21420	3.03150	3.96200
130 deg	0.25250	1.02140	2.16460	3.02500	3.29350
140 deg	0.21858	0.95361	2.12410	3.03520	3.89080
150 deg	0.19607	0.89560	2.09100	3.04500	3.86400
160 deg	0.16623	0.84410	2.06270	3.05380	3.84110
170 deg	0.14680	0.79956	2.03930	3.06160	3.82160
180 deg	0.12975	0.76007	2.01880	3.06890	3.80460

Table 7.4. Variation of the first five resonant frequency factors with respect to the panel angles for a circumferential number $n = 4$, $H/R = 0.3$.

ψ	Resonant Frequency Factors (βH)				
30 deg	3.09420	3.75470	5.05930	5.90670	6.76580
40 deg	2.26430	3.09475	4.20760	4.84595	5.79590
50 deg	1.72818	2.77095	3.59315	4.12775	5.30830
60 deg	1.36795	2.52823	3.22563	3.63270	4.97320
70 deg	1.11017	2.28725	3.04126	3.29160	4.72990
80 deg	0.91808	2.06083	2.92810	3.07662	4.54470
90 deg	0.77104	1.86397	2.78845	2.99652	4.40600
100 deg	0.60409	1.69730	2.65272	2.97665	4.29790
110 deg	0.56310	1.55640	2.54030	2.97430	4.21180
120 deg	0.48727	1.43650	2.44855	2.97865	4.14180
130 deg	0.42560	1.33360	2.37310	2.98640	4.08470
140 deg	0.37398	1.24510	2.31040	2.99520	4.03690
150 deg	0.33045	1.16750	2.25820	3.00450	3.99620
160 deg	0.29356	1.09990	2.21360	3.01330	3.96210
170 deg	0.26225	1.03990	2.17570	3.02210	3.93240
180 deg	0.23451	0.98660	2.14370	3.03040	3.90650

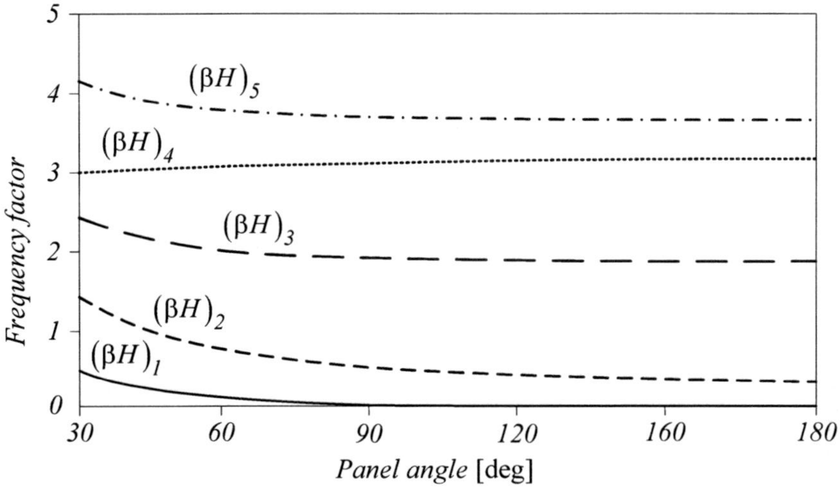

FIGURE 7.5. The variation of the first five resonant frequency factors versus panel angle for $n = 1$, $H/R = 0.3$, $\eta = 0.0$.

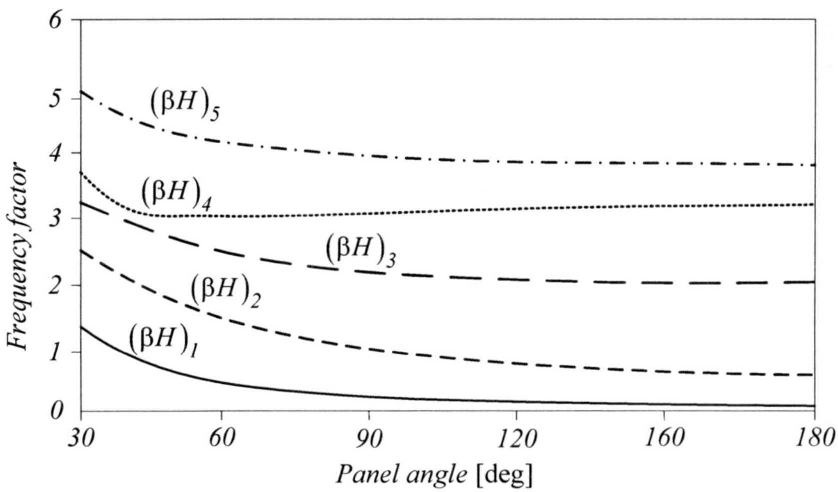

FIGURE 7.6. The variation of the first five resonant frequency factors versus panel angle for $n = 2$, $H/R = 0.3$, $\eta = 0.0$.

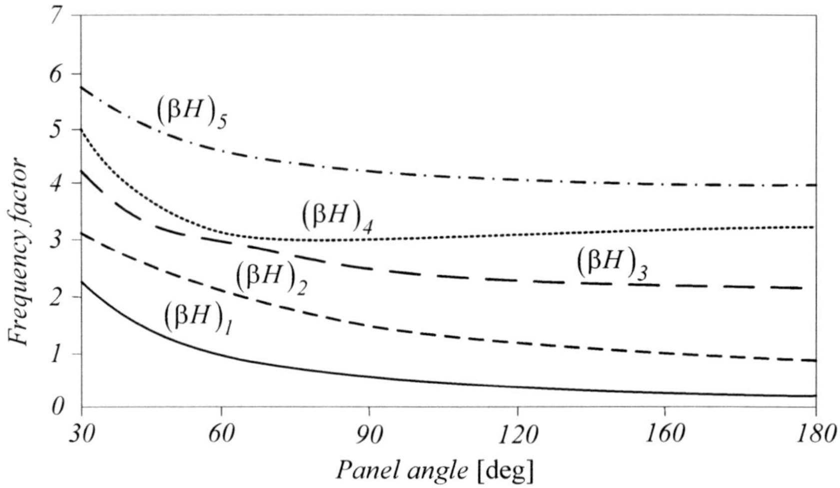

FIGURE 7.7. The variation of the first five resonant frequency factors versus panel angle for $n = 3$, $H/R = 0.3$, $\eta = 0.0$.

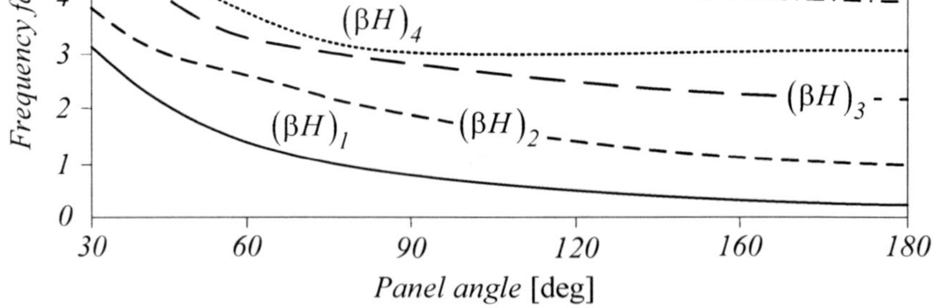

FIGURE 7.8. The variation of the first five resonant frequency factors versus panel angle for $n = 4$, $H/R = 0.3$, $\eta = 0.0$.

Table 7.5. Fundamental first resonant frequency factors for panel angle of 40°.

	n			
H/R	1	2	3	4
0.1	0.04050	0.17655	0.36278	0.57943
0.2	0.15420	0.53253	0.98784	1.45110
0.3	0.29370	0.91823	1.58630	2.26430
0.4	0.44377	1.28790	2.21282	2.93860
0.5	0.59413	1.62570	2.59250	3.45710
0.6	0.74008	1.93590	2.99570	3.89750
0.7	0.87938	2.18010	3.27500	4.28480
0.8	1.01100	2.39340	3.54750	4.62530
0.9	1.13400	2.60000	3.78000	4.92860
1.0	1.24740	2.74000	3.99270	5.20300

Table 7.6. Fundamental first resonant frequency factors for panel angle of 60°.

	n			
H/R	1	2	3	4
0.1	0.13898	0.08124	0.17643	0.29626
0.2	0.06554	0.26629	0.53286	0.82986
0.3	0.12974	0.48786	0.91846	1.36795
0.4	0.20309	0.71479	1.28760	1.81601
0.5	0.28109	0.93457	1.62690	2.27140
0.6	0.35896	1.14170	1.92880	2.64020
0.7	0.43672	1.33210	2.18910	2.93430
0.8	0.51198	1.50570	2.40920	3.17430
0.9	0.58422	1.67140	2.59500	3.39510
1.0	0.65360	1.81430	2.75400	3.58450

Table 7.7. Fundamental first resonant frequency factors for panel angle of 80°.

	n			
H/R	1	2	3	4
0.1	0.01041	0.04450	0.10234	0.17714
0.2	0.01347	0.15400	0.32840	0.53284
0.3	0.06596	0.29372	0.59160	0.91808
0.4	0.10523	0.44373	0.85596	1.28730
0.5	0.14799	1.62570	1.09798	1.6088
0.6	0.19196	0.74008	1.33980	1.92870
0.7	0.23627	0.87941	1.55640	2.18910
0.8	0.28049	1.0107	1.74880	2.40930
0.9	0.32399	1.1337	1.91920	2.59500
1.0	0.36599	1.24510	2.06600	2.75440

Table 7.8. Fundamental first resonant frequency factors for panel angle of 100°.

	n			
H/R	1	2	3	4
0.1	0.00829	0.02799	0.06558	0.11586
0.2	0.01970	0.09772	0.21891	0.36723
0.3	0.03460	0.19069	0.40750	0.60409
0.4	0.05629	0.29409	0.60407	0.94157
0.5	0.08041	0.40090	0.79661	1.21280
0.6	0.10570	0.50676	0.98038	1.46230
0.7	0.13160	0.60967	1.15260	1.68820
0.8	0.15760	0.70859	1.31140	1.88810
0.9	0.18330	0.80281	1.45841	2.06220
1.0	0.20850	0.89199	1.59110	2.21220

Table 7.9. Fundamental first resonant frequency factors for panel angle of 120°.

	n			
H/R	1	2	3	4
0.1	0.00352	0.01389	0.04494	0.08105
0.2	0.01040	0.06547	0.15318	0.26605
0.3	0.01870	0.12979	0.29371	0.48727
0.4	0.03060	0.20340	0.44374	0.671507
0.5	0.04550	0.27983	0.59356	0.93424
0.6	0.05799	0.35779	0.73949	1.10300
0.7	0.07234	0.43610	0.87999	1.33330
0.8	0.08719	0.51195	1.01100	1.50820
0.9	0.10161	0.58427	1.13320	1.66700
1.0	0.11621	0.65428	1.24690	1.80930

Table 7.10. Fundamental first resonant frequency factors for panel angle of 140°.

	n			
H/R	1	2	3	4
0.1	0.00210	0.01145	0.03146	0.05942
0.2	0.00525	0.02777	0.11235	0.19885
0.3	0.00974	0.09175	0.21858	0.37398
0.4	0.01551	0.14378	0.33571	0.55784
0.5	0.02200	0.20144	0.45428	0.73856
0.6	0.028750	.25975	0.57285	0.91025
0.7	0.36250	0.31967	0.68627	1.07350
0.8	0.04375	0.37739	0.79427	1.22560
0.9	0.05125	0.43252	0.89713	1.36570
1.0	0.05875	0.48776	0.99500	1.49420

Table 7.11. Fundamental first resonant frequency factors for panel angle of 160°.

	n			
H/R	1	2	3	4
0.1	0.00150	0.01007	0.02471	0.04509
0.2	0.00174	0.01350	0.08451	0.15427
0.3	0.00390	0.06625	0.16623	0.29356
0.4	0.00628	0.10508	0.25985	0.44354
0.5	0.00923	0.14742	0.35450	0.59426
0.6	0.01191	0.19142	0.45128	0.74026
0.7	0.22407	0.23714	0.54356	0.87996
0.8	0.01264	0.28000	0.63485	1.01000
0.9	0.01801	0.32285	0.71992	1.13420
1.0	0T02300	0.36688	0.80257	1.24800

Table 7.12. Fundamental first resonant frequency factors for panel angle of 180°.

	n			
H/R	1	2	3	4
0.1	0.000	0.00768	0.01128	0.03499
0.2	0.000	0.00925	0.06764	0.12268
0.3	0.000	0.04785	0.12975	0.23451
0.4	0.000	0.07704	0.20374	0.35908
0.5	0.000	0.10927	0.28114	0.48539
0.6	0.000	0.14261	0.35999	0.60899
0.7	0.000	0.17635	0.43714	0.72944
0.8	0.000	0.21030	0.51199	0.84325
0.9	0.000	0.24424	0.58406	0.94969
1.0	0.000	0.27639	0.65256	1.50140

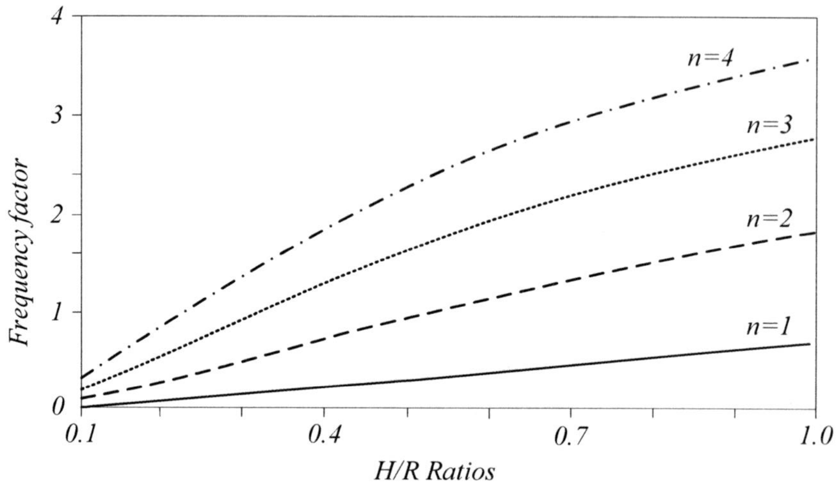

FIGURE 7.9. The variation of the fundamental frequency factors versus the ratio H/R for $n = 1$, 2, 3, 4, and panel angle of 40 deg.

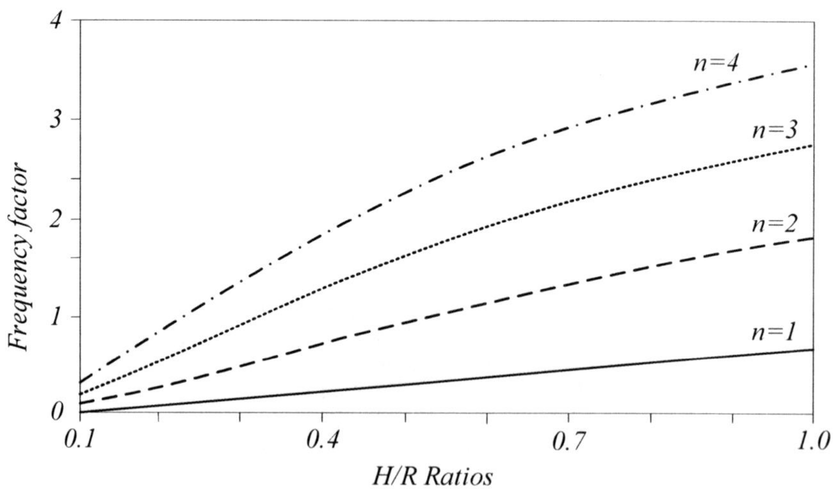

FIGURE 7.10. The variation of the fundamental frequency factors versus the ratio H/R for $n = 1$, 2, 3, 4, and panel angle of 60 deg.

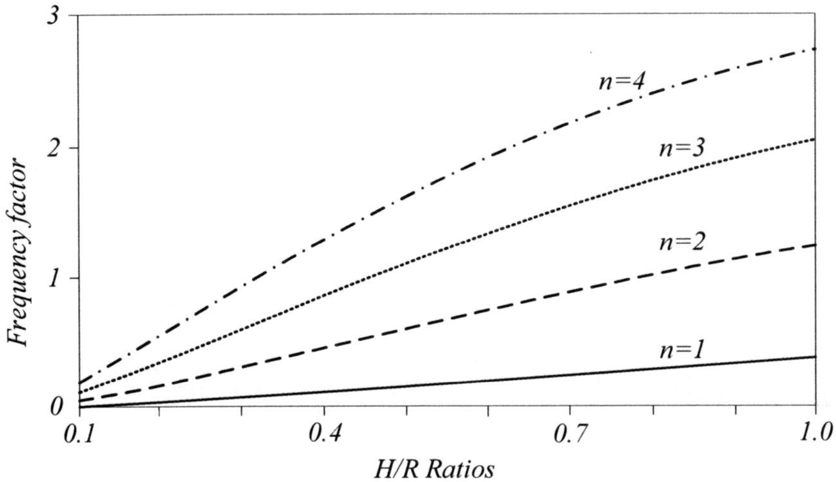

FIGURE 7.11. The variation of the fundamental frequency factors versus the ratio H/R for $n = 1$, 2, 3, 4, and panel angle of 80 deg.

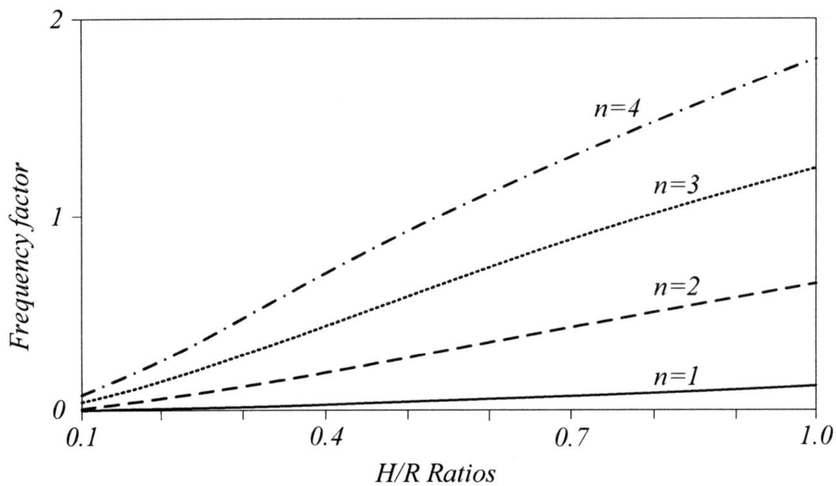

FIGURE 7.12. The variation of the fundamental frequency factors versus the ratio H/R for $n = 1$, 2, 3, 4, and panel angle of 100 deg.

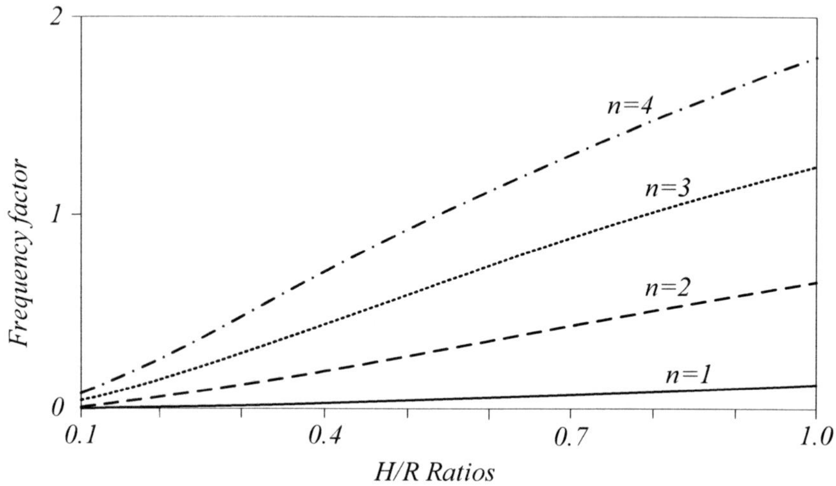

FIGURE 7.13. The variation of the fundamental frequency factors versus the ratio H/R for $n = 1, 2, 3, 4$, and panel angle of 120 deg.

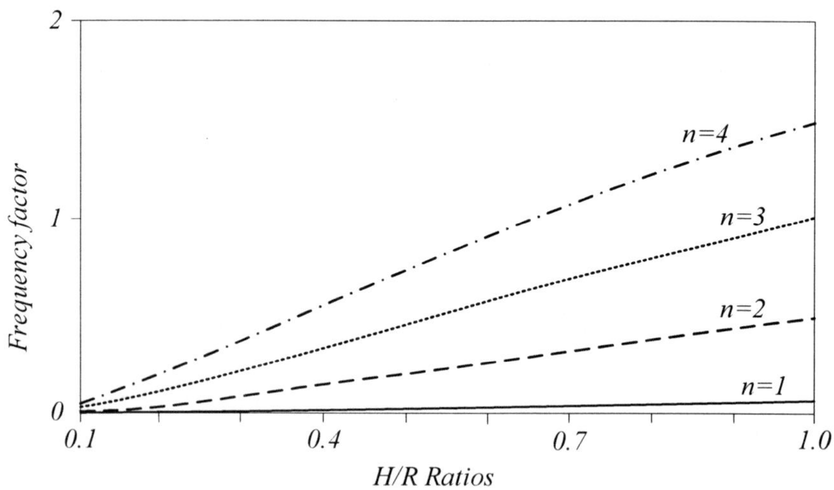

FIGURE 7.14. The variation of the fundamental frequency factors versus the ratio H/R for $n = 1, 2, 3, 4$, and panel angle of 140 deg.

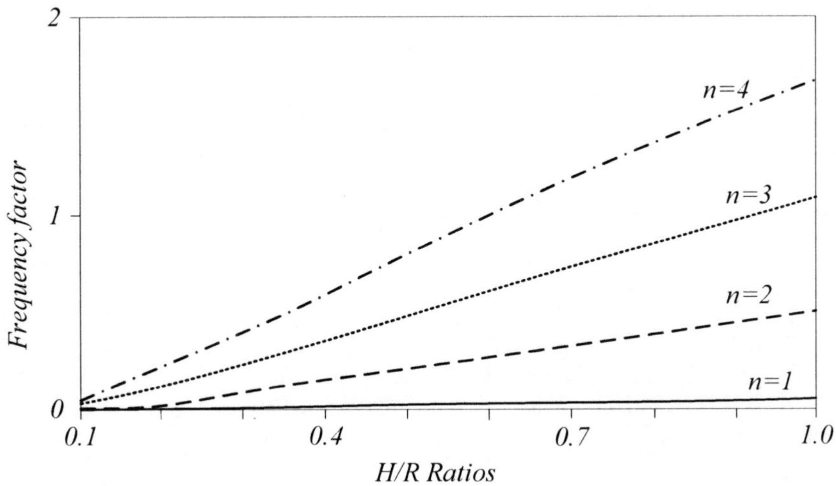

FIGURE 7.15. The variation of the fundamental frequency factors versus the ratio H/R for $n = 1, 2, 3, 4$, and panel angle of 160 deg.

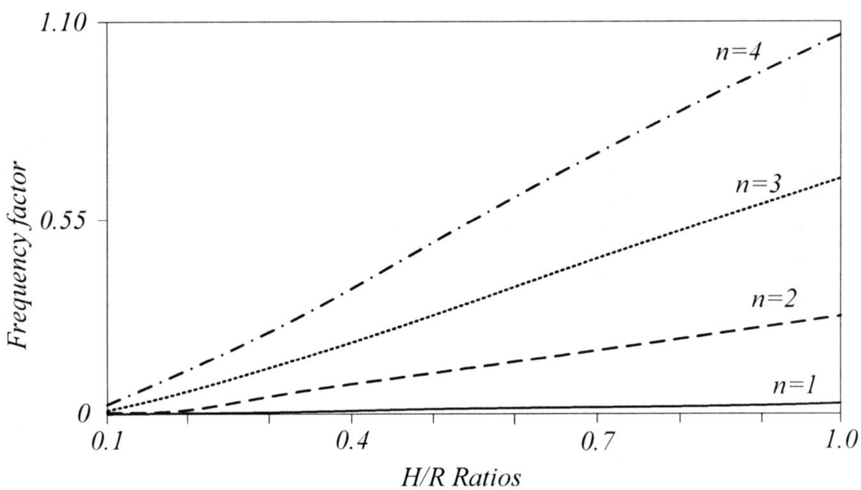

FIGURE 7.16. The variation of the fundamental frequency factors versus the ratio H/R for $n = 1, 2, 3, 4$, and panel angle of 180 deg.

Table 7.13. The resonant frequency factors and their respective modal loss-factors for $n = 1$ a panel with the angle of 60°.

		Mode 1		Mode 2		Mode 3	
H/R	η	βH	loss-factor	βH	loss-factor	βH	loss-factor
0.2	0.0	0.06555	0.00000	0.53631	0.00000	1.9142	0.0000
	.05	0.0634	0.04828	0.53633	0.04928	1.91443	0.04928
	0.1	0.0623	0.11040	0.54697	0.14479	1.96570	0.05835
0.4	0.0	0.2036	0.00000	0.95501	0.00000	2.14610	0.00000
	.05	0.2036	0.05070	0.97035	0.10568	2.14580	0.04848
	0.1	0.2036	0.09653	0.97035	0.12632	2.14415	0.08726
0.6	0.0	0.3592	0.00000	1.27678	0.00000	2.44270	0.00000
	.05	0.3594	0.05340	1.28041	0.09536	2.44104	0.07434
	0.1	0.3592	0.09810	1.27876	0.10105	2.44130	0.09727
0.8	0.0	0.5117	0.00000	1.49850	0.00000	2.75810	0.00000
	.05	0.5126	0.05180	1.49760	0.03976	2.74333	0.05111
	0.1	0.5113	0.09999	1.49852	0.09595	2.75899	0.08048
1.0	0.0	0.6535	0.00000	1.66920	0.00000	2.96430	0.00000
	.05	0.6538	0.05022	1.66880	0.06544	2.96340	0.04838
	0.1	0.6539	0.09622	1.67187	0.09643	2.96374	0.09811

Table 7.14. The resonant frequency factors and their respective modal loss-factors for $n = 2$ a panel with the angle of 60°.

		Mode 1		Mode 2		Mode 3	
H/R	η	βH	loss-factor	βH	loss-factor	βH	loss-factor
0.2	0.0	0.2660	0.000	1.0249	0.0000	2.0315	0.0000
	.05	0.2664	0.049	1.0246	0.0508	2.0319	0.0714
	0.1	0.2660	0.096	1.0257	0.0930	2.9283	0.1292
0.4	0.0	0.7145	0.000	1.7758	0.0000	2.7739	0.0000
	.05	0.7150	0.050	1.7705	0.0279	2.6142	0.0744
	0.1	0.7150	0.076	1.6606	0.1046	2.6132	0.0780
0.6	0.0	1.1415	0.000	2.2608	0.0000	3.0464	0.0000
	.05	1.1410	0.037	2.0466	0.0972	3.0405	0.0496
	0.1	1.1419	0.104	2.0494	0.1107	3.0471	0.1007
0.8	0.0	1.5093	0.000	2.5825	0.0000	3.2568	0.0000
	.05	1.5090	0.047	2.3199	0.0956	3.1927	0.0242
	0.1	1.5090	0.097	2.3373	0.1336	3.2499	0.0261
1.0	0.0	1.8063	0.000	2.8495	0.0000	3.5028	0.0000
	.05	1.8049	0.041	2.7487	0.0633	3.2685	0.0703
	0.1	1.8080	0.101	2.7054	0.1076	3.1728	0.2295

Table 7.15. The resonant frequency factors and their respective modal loss-factors for $n = 3$ a panel with the angle of 60°.

		Mode 1		Mode 2		Mode 3	
H/R	η	βH	loss-factor	βH	loss-factor	βH	loss-factor
0.2	0.0	0.53600	0.00000	1.50710	0.00000	2.45870	0.00000
	.05	0.54169	0.19658	1.52191	0.08756	2.35810	0.07100
	0.1	0.53344	0.09978	1.51110	0.12659	2.35810	0.09748
0.4	0.0	1.2883	0.00000	2.4457	0.0000	3.1380	0.00000
	.05	1.28850	0.04356	2.31481	0.09345	3.14080	0.04876
	0.1	1.2875	0.10155	2.31353	0.11943	3.14296	0.09774
0.6	0.0	1.9280	0.00000	2.9220	0.00000	3.6920	0.00000
	.05	1.95332	0.05080	2.92266	0.04389	3.6881	0.04417
	0.1	1.92322	0.09208	2.91995	0.09566	3.70665	0.08966
0.8	0.0	2.4077	0.00000	3.4107	0.0000	4.2057	0.00000
	.05	2.08945	0.18062	3.41048	0.04820	4.21067	0.05005
	0.1	2.08945	0.14903	3.21795	0.12370	4.23272	0.08328
1.0	0.0	2.7530	0.00000	3.9567	0.0000	4.6543	0.00000
	.05	2.36696	0.13445	3.82520	0.06640	4.65993	0.046231
	0.1	1.82193	4.14910	3.95179	0.10029	4.63600	0.08000

Table 7.16. The resonant frequency factors and their respective modal loss-factors for $n = 4$ a panel with the angle of 60°.

		Mode 1		Mode 2		Mode 3	
H/R	η	βH	loss-factor	βH	loss-factor	βH	loss-factor
0.2	0.0	0.83300	0.00000	1.95500	0.00000	2.81300	0.00000
	.05	0.74985	0.17840	1.8344	0.08798	2.71759	0.05973
	0.1	0.85297	0.25840	1.79890	0.16400	2.44725	0.59990
0.4	0.0	1.8600	0.0000	2.8533	0.0000	3.6933	0.0000
	.05	1.6353	0.4339	2.8544	0.04929	3.7031	0.0518
	0.1	1.6365	0.7997	2.8544	0.09547	3.7032	0.0816
0.6	0.0	2.6420	0.0000	3.5440	0.0000	4.5340	0.0000
	.05	2.2721	0.1115	3.5444	0.04619	4.5333	0.0352
	0.1	2.2711	0.0896	3.3358	0.11977	4.4081	0.0512
0.8	0.0	2.9750	0.0000	4.3877	0.0000	5.1927	0.0000
	.05	2.8547	0.3321	4.2471	0.07036	5.2174	0.0439
	0.1	2.8544	0.3566	4.3215	0.12564	5.1520	0.1256
1.0	0.0	3.5801	0.0000	5.1900	0.0000	5.8203	0.0000
	.05	3.4253	0.1263	5.2634	0.2356	5.8341	0.2568
	0.1	3.5478	0.5245	5.2235	0.3547	5.2360	0.2356

Table 7.17. The resonant frequency factors and their respective modal loss-factors for $n = 1$ a panel with the angle of 100°.

		Mode 1		Mode 2		Mode 3	
H/R	η	βH	loss-factor	βH	loss-factor	βH	loss-factor
0.2	0.0	0.01873	0.00000	0.35531	0.00000	1.85960	0.00000
	.05	0.01717	0.03352	0.35015	0.04921	1.86108	0.03912
	0.1	0.01748	0.08337	0.35016	0.09793	1.85431	0.01648
0.4	0.0	0.05081	0.00000	0.63003	0.00000	1.96360	0.00000
	.05	0.03956	1.61237	0.62897	0.04971	1.96473	0.07886
	0.1	0.03948	0.68782	0.62900	0.09631	1.93209	0.15389
0.6	0.0	0.1011	0.00000	0.84295	0.00000	2.1134	0.00000
	.05	0.09686	5.3437	0.87103	0.42735	2.1145	0.04652
	0.1	0.10668	0.09652	0.86977	0.27562	2.10806	0.08822
0.8	0.0	0.15611	0.00000	1.0078	0.00000	2.28510	0.00000
	.05	0.15733	0.04675	1.0267	0.21034	2.28258	0.04816
	0.1	0.15711	0.09784	1.0085	0.9836	2.26431	0.08805
1.0	0.0	0.21106	0.00000	1.1314	0.00000	2.42130	0.00000
	.05	0.20927	0.04952	1.0390	0.11890	2.42136	0.06119
	0.1	0.23180	2.33630	1.16654	0.25284	2.44801	0.05569

Table 7.18. The resonant frequency factors and their respective modal loss-factors for $n = 2$ a panel with the angle of 100°.

		Mode 1		Mode 2		Mode 3	
H/R	η	βH	loss-factor	βH	loss-factor	βH	loss-factor
0.2	0.0	0.09429	0.00000	0.63004	0.00000	1.94860	0.00000
	.05	0.09695	0.05165	0.58329	0.12933	1.96722	0.00371
	0.1	0.09635	0.09962	0.58329	1.1595	1.92471	0.07393
0.4	0.0	0.29436	0.00000	1.1249	0.00000	2.25463	0.00000
	.05	0.30967	1.22840	1.0339	0.12466	2.27094	0.00281
	0.1	0.30967	0.71119	1.0335	0.12657	2.27762	0.25814
0.6	0.0	0.50389	0.00000	1.4841	0.00000	2.63621	0.00000
	.05	0.53277	0.90880	1.48161	0.49642	2.51452	0.06692
	0.1	0.50706	0.1001	1.4773	0.7503	2.51301	0.10908
0.8	0.0	0.70748	0.00000	1.7469	0.00000	2.93472	0.00000
	.05	0.54536	1.0573	1.6461	0.04251	2.93961	4.49877
	0.1	0.72071	0.19982	1.6458	0.8968	2.96422	0.13912
1.0	0.0	0.89649	0.00000	1.9359	0.00000	2.98879	0.00000
	.05	0.73001	0.52668	1.81884	0.10439	2.99361	0.05174
	0.1	0.73080	0.37482	1.8189	0.11930	2.83130	0.10928

Table 7.19. The resonant frequency factors and their respective modal loss-factors for $n = 3$ a panel with the angle of 100°.

		Mode 1		Mode 2		Mode 3	
H/R	η	βH	loss-factor	βH	loss-factor	βH	loss-factor
0.2	0.0	0.22150	0.00000	0.92340	0.00000	2.08430	0.00000
	.05	0.21929	0.05023	0.92670	0.48678	2.09288	0.04960
	0.1	0.21857	0.09784	1.00357	0.47030	2.09740	0.09815
0.4	0.0	0.59949	0.00000	1.61190	0.00000	2.63770	0.00000
	.05	0.62341	0.46938	1.50570	0.83790	2.68410	0.09857
	0.1	0.60454	0.09925	1.61800	0.09331	2.60395	0.10031
0.6	0.0	0.95760	0.00000	2.08430	0.00000	2.98870	0.00000
	.05	1.00189	0.2472	2.08380	0.04895	2.98671	0.04738
	0.1	0.98000	0.09615	2.08805	0.10336	2.98990	0.11902
0.8	0.0	1.31490	0.00000	2.39480	0.00000	3.12370	0.00000
	.05	1.33790	0.18609	2.27801	0.08490	3.13136	0.05765
	0.1	1.33720	1.69876	2.27747	0.11387	3.11114	0.07605
1.0	0.0	1.58490	0.00000	2.63787	0.00000	3.31270	0.00000
	.05	1.59189	0.04997	2.66218	0.06380	3.31339	0.05072
	0.1	1.27360	0.27786	2.66194	0.10538	3.32038	0.11931

Table 7.20. The resonant frequency factors and their respective modal loss-factors for $n = 4$ a panel with the angle of 100°.

		Mode 1		Mode 2		Mode 3	
H/R	η	βH	loss-factor	βH	loss-factor	βH	loss-factor
0.2	0.0	0.36140	0.00000	1.21810	0.00000	2.26370	0.00000
	.05	0.36855	0.04974	1.11960	0.16891	2.26573	0.04831
	0.1	0.36788	0.09967	1.23475	0.0993	2.26162	0.09486
0.4	0.0	0.93698	0.00000	2.07080	0.00000	2.96170	0.00000
	.05	0.94195	0.04824	2.08830	0.07365	2.95599	0.05012
	0.1	0.95890	0.19179	1.94711	0.11596	2.97320	0.12119
0.6	0.0	1.45890	0.00000	2.56580	0.00000	3.24070	0.00000
	.05	1.27540	0.20212	2.57890	0.06299	3.25010	0.05276
	0.1	1.27530	0.23354	2.44150	0.11499	3.24002	0.04989
0.8	0.0	1.89150	0.00000	2.90801	0.00000	3.59960	0.00000
	.05	1.63720	0.17180	2.92280	0.05774	3.61222	0.59680
	0.1	1.63720	0.11780	2.79260	0.11781	3.61205	0.63160
1.0	0.0	2.21510	0.00000	3.27340	0.00000	3.93970	0.00000
	.05	1.98771	0.10534	3.15024	0.07199	3.95546	0.49920
	0.1	1.98770	0.11954	3.15024	0.11441	3.95564	0.62510

Table 7.21. The resonant frequency factors and their respective modal loss-factors for $n = 1$ a panel with the angle of 140°.

		Mode 1		Mode 2		Mode 3	
H/R	η	βH	loss-factor	βH	loss-factor	βH	loss-factor
0.2	0.0	0.00524	0.00000	0.27400	0.00000	1.83990	0.00000
	.05	0.00543	0.2779	0.27783	0.04975	1.81740	0.03713
	0.1	0.00522	0.01378	0.23310	0.49152	1.86632	0.09009
0.4	0.0	0.01554	0.00000	0.49955	0.00000	1.91720	0.00000
	.05	0.01553	0.02047	0.52652	1.18440	1.77209	0.03385
	0.1	0.15841	0.97883	0.52642	0.65396	1.77209	0.06542
0.6	0.0	0.02875	0.00000	0.67999	0.00000	2.00740	0.00000
	.05	0.02936	0.04586	0.67726	0.04924	1.86310	0.03385
	0.1	0.02937	0.09151	0.59970	0.25452	1.86590	0.01308
0.8	0.0	0.04370	0.00000	0.80887	0.00000	2.12340	0.00000
	.05	0.04399	0.04956	0.81477	0.05017	1.98861	0.07114
	0.1	0.05202	14.2186	0.81443	0.09936	1.98790	0.06618
1.0	0.0	0.05875	0.00000	0.92487	0.00000	2.23940	0.00000
	.05	0.05976	0.04737	0.92283	0.04970	2.24130	0.04714
	0.1	0.05942	0.04266	0.92283	0.09894	2.29502	0.08720

Table 7.22. The resonant frequency factors and their respective modal loss-factors for $n = 2$ a panel with the angle of 140°.

		Mode 1		Mode 2		Mode 3	
H/R	η	βH	loss-factor	βH	loss-factor	βH	loss-factor
0.2	0.0	0.02771	0.00000	0.46810	0.00000	1.89470	0.00000
	.05	0.03915	9.05458	0.41359	0.37977	1.73720	0.07914
	0.1	0.03915	5.75293	0.41313	0.24430	1.73750	0.04916
0.4	0.0	0.14089	0.00000	0.83458	0.00000	2.07790	0.00000
	.05	0.14493	0.05046	0.75188	0.27593	1.94350	0.08873
	0.1	0.14446	0.09894	0.83703	0.09399	1.94350	0.08773
0.6	0.0	0.25869	0.00000	1.10940	0.00000	2.31350	0.00000
	.05	0.26085	1.04995	1.01266	2.22947	2.18696	0.08118
	0.1	0.26036	0.10005	1.11501	0.08570	2.33310	0.36700
0.8	0.0	0.37648	0.00000	1.31880	0.00000	2.57520	0.00000
	.05	0.37715	0.04961	1.32220	0.04570	2.44066	0.08600
	0.1	0.37715	0.09974	1.32340	0.09209	2.44090	0.11498
1.0	0.0	0.48120	0.00000	1.47590	0.00000	2.83700	0.00000
	.05	0.48653	0.05004	1.47915	0.04841	2.84351	0.15302
	0.1	0.48724	0.09985	1.47898	0.09228	2.98607	0.11933

Table 7.23. The resonant frequency factors and their respective modal loss-factors for $n = 3$ a panel with the angle of 140°.

		Mode 1		Mode 2		Mode 3	
H/R	η	βH	loss-factor	βH	loss-factor	βH	loss-factor
0.2	0.0	0.11471	0.00000	0.67752	0.00000	1.97320	0.00000
	.05	0.11324	0.04878	0.67322	0.04888	1.96880	0.04622
	0.1	0.11335	0.09940	0.67366	0.09328	1.84320	0.07692
0.4	0.0	0.33722	0.00000	1.20100	0.0000	2.31340	0.00000
	.05	0.28127	0.64600	1.10305	01.8709	2.18457	0.07756
	0.1	0.28125	0.52161	1.19550	0.08721	2.19170	0.31956
0.6	0.0	0.57688	0.00000	1.56920	0.0000	2.98680	0.00000
	.05	0.46265	1.02314	1.46360	0.11117	2.59470	0.24969
	0.1	0.58360	0.23266	1.46270	0.09801	2.93900	0.07726
0.8	0.0	0.79599	0.00000	1.83990	0.0000	2.96100	0.00000
	.05	0.80639	0.19225	1.73170	0.08463	2.95980	0.04353
	0.1	0.55671	0.85157	1.61790	0.10256	2.95627	0.08946
1.0	0.0	0.98932	0.00000	2.04610	0.0000	3.01260	0.00000
	.05	1.01250	0.20017	1.93720	0.07510	2.90569	0.07337
	0.1	0.82483	0.32240	1.93720	0.10642	2.90515	0.14827

Table 7.24. The resonant frequency factors and their respective modal loss-factors for $n = 4$ a panel with the angle of 140°.

		Mode 1		Mode 2		Mode 3	
H/R	η	βH	loss-factor	βH	loss-factor	βH	loss-factor
0.2	0.0	0.20311	0.00000	0.88620	0.00000	2.07180	0.00000
	.05	0.17245	0.91046	0.80918	0.14490	1.94560	0.08357
	0.1	0.17242	0.54373	0.91065	1.81566	1.94560	0.09006
0.4	0.0	0.55110	0.00000	1.54340	0.00000	2.58720	0.00000
	.05	0.55773	0.04929	1.44716	0.08357	2.60207	0.06695
	0.1	0.55773	0.10138	1.44716	0.10379	2.47430	0.10554
0.6	0.0	0.91198	0.00000	2.00740	0.00000	2.96100	0.00000
	.05	0.73497	0.59900	1.88220	0.09163	2.85400	0.07753
	0.1	0.73497	0.41465	2.31670	0.11291	2.98200	0.09850
0.8	0.0	1.22130	0.00000	2.19246	0.00000	3.08990	0.00000
	.05	1.09381	0.16901	2.19180	0.09192	3.08711	0.04708
	0.1	1.25230	0.18195	2.19188	0.18950	3.08338	0.08045
1.0	0.0	1.49190	0.00000	2.54860	0.00000	3.25574	0.00000
	.05	1.27600	0.17009	2.43800	0.08212	3.25695	0.06051
	0.1	1.20609	0.18162	2.43808	0.10989	3.25727	0.08005

Table 7.25. The resonant frequency factors and their respective modal loss-factors for $n = 1$ a panel with the angle of 180°.

		Mode 1		Mode 2		Mode 3	
H/R	η	βH	loss-factor	βH	loss-factor	βH	loss-factor
0.2	0.0	0.24560	0.00000	3.13790	0.00000	3.64830	0.00000
	.05	0.23659	7.62900	2.83715	0.14100	6.18061	0.06130
	0.1	0.23659	3.86220	2.83719	0.14273	6.18062	0.04705
0.4	0.0	0.44193	0.00000	3.13792	0.0000	3.68762	0.00000
	.05	0.39112	0.37310	2.98101	0.09500	6.13708	0.04360
	0.1	0.39112	0.27106	2.98102	0.10388	6.13734	0.06039
0.6	0.0	0.59899	0.00000	1.96011	0.0000	3.13795	0.00000
	.05	0.46469	0.60110	2.99236	0.08181	3.15551	0.06085
	0.1	0.46370	0.00000	2.99236	0.11856	3.15552	0.10710
0.8	0.0	0.71679	0.00000	2.05172	0.0000	3.13799	0.00000
	.05	0.74308	0.47026	3.01786	0.08020	3.15374	0.05595
	0.1	0.74308	0.31770	3.01778	0.1009	3.13764	0.08352
1.0	0.0	0.82151	0.00000	2.14336	0.0000	3.15115	0.00000
	.05	0.73238	0.21723	1.99775	0.0804	3.14551	0.05038
	0.1	0.73243	0.21723	1.99761	0.0803	3.03921	0.06224

Table 7.26. The resonant frequency factors and their respective modal loss-factors for $n = 2$ a panel with the angle of 180°.

		Mode 1		Mode 2		Mode 3	
H/R	η	βH	loss-factor	βH	loss-factor	βH	loss-factor
0.2	0.0	0.00925	0.00000	0.37648	0.00000	3.12480	0.00000
	.05	0.00837	0.79297	0.33730	0.26560	3.12360	0.04224
	0.1	0.00895	0.43765	0.33705	0.20342	2.85360	0.12411
0.4	0.0	0.07544	0.00000	0.67752	0.00000	1.98630	0.00000
	.05	0.07737	0.04918	0.61386	0.20377	2.97812	0.06603
	0.1	0.07758	0.09505	0.61386	0.17855	2.97851	0.10901
0.6	0.0	0.14089	0.00000	0.91311	0.00000	2.15640	0.00000
	.05	0.14314	0.48299	0.82599	0.10406	2.17252	0.39926
	0.1	0.16446	4.06440	0.82559	0.17829	2.96210	0.11287
0.8	0.0	0.20633	0.00000	1.08330	0.00000	2.35271	0.00000
	.05	0.17271	1.24360	0.99068	0.17895	2.96492	0.62076
	0.1	0.17271	0.72590	0.99022	0.16324	2.96501	0.12271
1.0	0.0	0.27178	0.00000	1.22722	0.00000	2.56210	0.00000
	.05	0.24465	0.43486	1.12660	0.12290	1.24856	0.27447
	0.1	0.24465	0.30849	1.12483	0.13562	1.24841	0.23229

Table 7.27. The resonant frequency factors and their respective modal loss-factors for $n = 3$ a panel with the angle of 180°.

		Mode 1		Mode 2		Mode 3	
H/R	η	βH	loss-factor	βH	loss-factor	βH	loss-factor
0.2	0.0	0.67640	0.00000	0.53355	0.00000	1.90770	0.00000
	.05	0.71216	4.9148	0.45682	0.56638	2.83280	0.10510
	0.1	0.07121	2.4360	0.55262	0.29234	2.83320	0.10756
0.4	0.0	0.20633	0.00000	0.95237	0.00000	2.14330	0.00000
	.05	0.21687	2.0449	0.86104	0.26039	2.15770	0.36820
	0.1	0.21687	1.11111	0.86104	1.91680	2.94460	0.10787
0.6	0.0	0.36340	0.00000	1.26650	0.00000	2.44430	0.00000
	.05	0.31272	0.56109	1.16493	0.16195	2.46601	0.24310
	0.1	0.31282	0.37511	1.16427	0.14875	2.93520	0.09212
0.8	0.0	0.50737	0.00000	1.50260	0.00000	2.75840	0.00000
	.05	0.45243	0.28587	1.37960	0.16027	2.76640	0.00296
	0.1	0.45243	0.23574	1.38001	0.13439	2.62101	0.11333
1.0	0.0	0.65130	0.00000	1.67220	0.00000	2.96780	0.00000
	.05	0.58754	0.25118	1.51290	0.16950	2.78439	0.07971
	0.1	0.58754	0.21518	1.51290	0.13631	2.78346	0.12081

Table 7.28. The resonant frequency factors and their respective modal loss-factors for $n = 4$ a panel with the angle of 180°.

		Mode 1		Mode 2		Mode 3	
H/R	η	βH	loss-factor	βH	loss-factor	βH	loss-factor
0.2	0.0	0.11471	0.00000	0.69021	0.00000	2.33960	0.00000
	.05	0.12227	0.05050	0.63066	0.20859	2.9610	0.0549
	0.1	0.09551	1.31630	0.71244	0.15657	2.9610	0.5498
0.4	0.0	0.36340	0.00000	1.24030	0.00000	2.33960	0.00000
	.05	0.38232	0.80869	1.12140	0.14215	2.20672	0.10207
	0.1	0.38232	1.45073	1.12140	0.17150	2.20406	0.08679
0.6	0.0	0.61555	0.00000	1.62080	0.00000	2.75481	0.00000
	.05	0.55147	0.16502	1.51211	0.10487	2.75251	0.04422
	0.1	0.55147	0.16973	1.51211	0.11771	2.62370	0.10725
0.8	0.0	0.84754	0.00000	1.90430	0.00000	2.97390	0.00000
	.05	0.84433	0.07768	2.84400	0.07768	2.98016	0.06370
	0.1	0.84433	0.10220	2.84463	0.07590	2.97961	0.09459
1.0	0.0	1.05380	0.00000	2.11050	0.00000	3.02255	0.00000
	.05	0.95712	0.21740	2.92687	0.06823	3.04865	0.07492
	0.1	0.95712	0.22030	2.11276	0.29441	4.68622	0.07238

Tables 7.13 to 7.28 are the frequency factors and loss-factors computed for different panel angles and different material damping factors. The panel

angle is varying form 30 deg to 180 deg and the loss-factors changed proportionately to damping factors. The damping factor of 0.05 and 0.1 were considered and the effect of damping to reduce the amplitude at resonance is shown in the graphs of modal response.

7.7 Finite Element Model

In this section the plane strain Finite Element Analysis of thick cylindrical panel is presented. The details of modeling and input are described. Natural frequencies and mode shapes for free vibrations of different size panels are computed. These results are nondimensionalized and are compared with those analytically determined in previous chapters.

Finite element method is used here to verify the values of frequency factors obtained from analytical method as described in chapters 4 and 5. The panel is modeled with same H/R ratio of 0.3, the panel is elastic with Elasticity modulus ($E = 30 \times 10^6\ psi$) and Poission's ratio of 0.25. Model consisted of six rows of elements in radial direction and fifty columns in tangential direction. Two-dimensional solid elements were considered. The boundary conditions are assumed by supporting the panel on rollers, free to move only in r direction and fixed in 0 directions.

The finite element package used is for the analysis was mechanical, which uses the Pversion of finite element method which uses iso-parametric element commonly known as higher order elements.

The order of polynomial used for approximation of the model can be selected from a minimum of 1 to maximum of 9. The relative percentage of convergence can be selected depending on the complexity of the model. While doing the modal analysis 0.01% convergence on modal frequency was specified. The model converged at 0.01% convergence and computed the first twenty resonant frequencies. The model with panel angle of 180°, 120° and 90° converged at maximum edge order of 5, 6 and 7, respectively.

The units of modal frequencies obtained from Finite element in Hertz were converted to by multiplying with a conversion factor. The analysis was repeated for different panel models for the angles of 180°, 120° and 90°. To know the accuracy of the results a model was made with a panel angle of 360° degrees, corresponding to a cylinder. The same procedure was used to compute the resonant frequencies. These results were compared with those of Armenakas, which are found to be in satisfactory agreement. Table 7.29 shows the comparison with Armenakas frequency factors, and

Tables 7.30 through 7.32 compare with those analytically determined in previous chapters.

Table 7.29. Comparison of the first six dimensionless frequency factors for the cylinder with $H/R = 0.3$.

FEM	Armenakas
0.035965	0.03596
0.095561	0.09556
0.163039	0.16306
0.169974	0.16997
0.225237	0.22524
0.351493	0.35149

Table 7.30. Comparison of the first six dimensionless frequency factors for the cylindrical panel with $H/R = 0.3$. Angle of panel = 180°.

Analytical Method	FEM
0.04841	0.04405
0.12975	0.12137
0.23480	0.23763
0.53975	0.53094

Table 7.31. Comparison of the first six dimensionless frequency factors for the cylindrical panel with $H/R = 0.3$. Angle of panel = 120°.

Analytical Method	FEM
0.12970	0.10482
0.29370	0.27353
0.48727	0.48567
0.76010	0.73383
1.43650	1.48597
1.88551	1.89861

Table 7.32. Comparison of the first six dimensionless frequency factors for the cylindrical panel with $H/R = 0.3$. Angle of panel = 90°.

Analytical Method	FEM
0.2347	0.20216
0.4878	0.4880
1.9216	1.9939
2.1437	2.1570

7.8 Concluding Remarks

Dynamic response of cylindrical panels excited by a harmonic force is determined. In this thesis accurate analysis is given to make possible the computation of resonant frequencies and corresponding vibration modes. The flexural, torsional and their corresponding thickness modes are computed. The analysis provides results for infinitely long cylinders with either elastic or visco-elastic properties.

The chapter presented the frequency factors to cover most practical conditions in the form of design tables. The results can be divided into three different design parameters. In the first case the first five resonant frequencies are computed for panels with panel angles ranging from 30° to 180° with 10° increment, with H/R of 0.3. In the second case the fundamental resonant frequencies for different H/R ratio ranging from 0.2 to 1.0 with 0.2 increment, for circumferential numbers 1 to 4 are covered. Finally the loss-factors are computed for different panels with angles of 30° to 180° with three structural damping factors of 0.0, 0.05 and 0.1. The H/R ratio is kept constant at 0.3.

The effect of visco-elastic material has been considered by including the complex moduli. The effect of damping on modal parameters has been investigated and results are provided for two different values of loss-factors for the visco-elastic medium.

The finite element comparison is made to verify the validity of the results obtained by analytical method. The developed analytical procedure is not limited to panels which are thin but is also applicable to thick panels. The comparison has indicated that the presented results have excellent agreement with computed resonant frequencies using Finite element method.

The problem of dynamics of thick cylindrical finite panels with different end conditions has not been widely considered. Analytical methods for vibrations of infinitely long or simply supported end conditions is established. However the theory should be extended to include the different panel angles and other types of boundary conditions.

Although, the effect of material damping in the medium has been considered here, the provided damping treatment is not as efficient as constrained layer damping. The developed analytical procedure can be adopted for constrained multi-layered cylindrical panels to determine the damped response.

7.9 Key Symbols

B	stress vector
E	Young's modulus
F_n, F_n^*	functions of Bessel functions
G, μ	shear modulus, modulus of rigidity
G_n, G_n^*	functions of Bessel functions
H_i	thickness of the layer i
H	thickness of the cylinder
H_n, H_n^*	functions of Bessel functions
$I_n,\ K_n$	modified Bessel functions
$J_n,\ Y_n$	Bessel functions
L	axial half wavelength
n	wave number
$n = 0$	breathing mode
$n = 1$	rigid body mode
$n > 1$	lobar modes
r	radial direction
$r = a$	inner surface boundary
$r = b$	outer surface boundary
R	mean radius of the cylinder
t	time
$[T]$	matrix of coefficients of Bessel functions
u	displacement
$u_r,\ u_\theta,\ u_z$	displacements in r, θ, z directions
v_1	velocity of propagation of the dilation wave
v_2	velocity of propagation of the distortional wave
W_n, Z_n	Bessel functions
X	vector of constants, vector of unknowns
z	axial direction
α	$= \sqrt{\rho p^2/G}$
β	$= \sqrt{\rho p^2/(\lambda + 2G)}$
γ	wave number in axial direction
ε	volumetric strain
η	loss-factor
θ	transverse direction
λ	Lame's elastic constant
μ, G	shear modulus, modulus of rigidity
ν	Poisson's ratio
ρ	density, circular frequency
σ	normal stress
$\sigma_{rr},\ \sigma_{\theta\theta},\ \sigma_{zz}$	direct stress in the r, θ, z directions

τ	shear stress
$\tau_{r\theta}$	shear stress along θ and perpendicular to r
τ_{rz}	shear stress along z and perpendicular to r
$\tau_{\theta z}$	shear stress along z and perpendicular to θ
ω	excitation frequency
ω_s	the lowest simple thickness shear frequency
ω_r, ω_θ, ω_z	elastic rotations in the r, θ, z directions
Ω	frequency factor
$\zeta = \pi/L$	axial wave number
ρ	density
σ	normal stress
σ_{rr}, $\sigma_{\theta\theta}$, σ_{zz}	direct stress in the r, θ, z directions
τ	shear stress
$\tau_{r\theta}$	shear stress along θ and perpendicular to r
τ_{rz}	shear stress along z and perpendicular to r
$\tau_{\theta z}$	shear stress along z and perpendicular to θ
$\zeta = \pi/L$	axial wave number
$\Im\omega$	imaginary part of ω
$\Re\omega$	real part of ω
∇	gradient

References

M. Abramowitz and I. Stegun. *Handbook of mathematical functions.* Dover publisher, 1965.

J. A. Agbasiere and P. Grootenhuis. Flexural vibration of symmetrical multi-layered beams with visco-elastic damping. *Journal of Mechanical Engineering: Science, 10 (3): 269-280, 1968.*

M. Amabili. *Nonlinear Vibrations and Stability of Shells and Plates.* Cambridge Press, 2008.

A. Armenakas, A., D. Gazis, and G. Herrmann. *Free vibrations of circular cylindrical shells.* Permagon Press Inc., 1969.

A. Armenakas. Propagation of harmonic waves in composite circular cylindrical shells. part ii: numerical results. *AIAA Journal, 9: 599-605, 1971.*

A. E. Armenakas. Propagation of harmonic waves in composite circular cylindrical shells. part I: theoretical investigation", *AIAA Journal, 5: 740-744, 1967.*

R. N. Arnold and G. B. Warburton. Flexural vibrations of the walls of thin cylindrical shells having freely supported ends. *Proceedings of the Royal Society, A 197: 238-256, 1949.*

R. N. Arnold, R.N. and G. B. Warburton. The flexural vibrations of thin cylinders. *Proceedings of the Institution of Mechanical Engineering Science. 167 A: 62-74, 1953.*

R. Beckett and J. Hurt. *Numerical calculations and algorithms.* McCraw-Hill Inc., 1967.

W. W. Bell. *Special functions for scientist and engineers*, van Nostrand, 1968.

L. A. Bergman, H. H. Hilton, and T. C. Tsao. The effect of viscoelasticity on the performance of reaction mass actuators. *Proceedings of damping' 93, GAB.* 1-8, 1993.

K. Bhaskar and T. K. Varadan. Analytical solution for a pinched laminated cylinder. *Journal of Applied Mechanics, 1082-1083, 1990.*

M. P. Bienieck and A. M. Freudenthal. Forced vibrations of cylindrical sandwich shells. *Journal of the Aerospace Sciences. 29:180-184, 1962.*

M. B. Braga, F. E. Barbone, and C. Herrmann. Wave propagation in fluid loaded laminated cylindrical shells. *Applied Mechanics Review, l43 (5) part 2: s359-s365, 1990.*

L. L. Bucciarelli, L.L. and N. Dworsky. *An essay in the history of the theory of e1asticity*, D. Reidel Publishing Co., Dordrecht, Holland, 1980.

J. T. Cannon and S. Dostrovsky. *The evolution of dynamics: vibration theory from 1687 to 1742.* Springer Verlag, 1981.

C. Chree. Longitudinal vibrations of a circular bar. *The Ouarterly Journal of Pure and Applied Mathematics, London 21:287-298, 1886.*

A. P. Christoforou and S. R. Swanson. Analysis of simply-supported

H.R. Hamidzadeh, R.N. Jazar, *Vibrations of Thick Cylindrical Structures*
DOI 10.1007/978-0-387-75591-5, © Springer Science+Business Media, LLC 2010

orthotropic cylindrical shells subjected to lateral impact loads. *Transaction of the ASME, 57: 376-382, 1990.* ???

S. Conte. *Elementary Numerical Analysis.* McGraw Hill, 1965.

S. Conte and DeBoor, *Numerical Analysis as Algorithmic Approach, Third Edition*, McGraw Hill. 1.980.

S. Crampin. The dispersion of surface waves in multilayered anisotropic media. *Geophys. J.R. Astr. Society,21:387-402, 1970.*

S. H. Crandall. The hysteretic damping model in vibration theory. *Journal of Mechanical Engineering Science, 205: 23-28, 1991.*

R. A. DiTaranto, Y. P. Lu, and B. E. Douglas. Forced response of a discontinuously constrained damped ring. *The Journal of the Acoustical Society of America. 54: 74-79, 1973.*

S. B. Dong. Freevibration of laminated orthotropic cylindrical shells. *Journal of the Acoustical Society of America, 44: 1628-1635, 1968.*

M. El-Raheb and P. Wagner. Damped response of shells by a constrained visco-elastic layer. *Journal of Applied Mechanics, 53(44): 902-908, 1986.*

D. J. Ewins. *Modal testing: theory and practice.* Research Studies Press Ltd, 1986.

W. Flugge. *Static und Dynamik der Schalen.* Berlin, Julius Springer, 1934.

W.Flugge. *Viscoelasticity.* Blaisdell Publishing Co., London, 1967.

H. Ford and J. M. Alexander. *Advanced mechanics of materials.* Longman Group Limited, London, 1963.

K. Forsberg. Influence of Boundary conditions on the modal characteristics of thin cylindrical shells. *AIAA Journal 2(12):2150-2157, 1964.*

Y. C. Fung and E. E. Sechier. Thin shell structures: Theory, experiment, and design. *NASA SP-221, 1972.*

D. C. Gazis. Exact analysis of plane-strain vibration of thick walled hollow cylinders. *Journal of the Acoustical Society of America 30:786-794. 1958.*

D. C. Gazis. Three dimensional investigation of the propagation of waves of hollow circular cylinders- I. Analytical Foundation, II. Numerical Results. *Journal of the Acoustical Society of America 31:568-578, 1959.*

B. M. L. Gladwell and. K. Vijay. Natural frequencies of free finite length circular cylinders. *Journal of Sound and Vibration 42:387-397, 1975.*

J. E. Greenspon. Flexural vibrations of a thick walled circular cylinder according to the exact theory of elasticity. *Journal of Aero/Space Science 27:1365-1373, 1957.*

P. Grootenhuis. The control of vibrations with visco-elastjc materials. *Journal of Sound and Vibration 1l(4): 42l-423, 1970.*

H. R. Hamidzadeh. Damping of Lobar Vibrations in Cylindrical Structures. *Proceedings of the 61st Shock and Vibration Symposium, 1: 233-240, 1990.*

H. R. Hamidzadeh. Flexural Vibrations of Thick Cylindrical Panels. *Proceedings of the ASME Design Engineering Technical Conferences, DETC97/VIB-*

4083, 1997.

H. R. Hamidzadeh. The effect of visco-elastic core thickness on modal loss factors of a thick three-layer cylinder. *Journal of Multi-body dynamics. 223(1): 1-8, 2009.*

H. R. Hamidzadeh and D. Chandler. Circumferential vibrations of three layered sandwich cylinders. *The Proceedings of the Thirteenth Biennial ASME Conference on Mechanical Vibrations and Noise, DE 36: 233- 237, 1991.*

H. R. Hamidzadeh and Y. Jiang. modal loss-factors of Thick Three-Layer Cylinders with Visco-elastic Core. *Proceedings of the ASME Design Engineering Technical Conferences, 3, Part C, 3-8, 1995.*

H. R. Hamidzadeh and A. A. Medhora. Free transverse vibrations of composite shells with discontinuous constraining layers. *ASME International Mechanical Engineering Congress, AD 46: 105-110, 1995.*

H. R. Hamidzadeh and L. C. Moxey. Dynamic models for components considered in the design of a solar concentrator. *Proceedings of the 3rd International Symposium on Multi-body Dynamics: Monitoring and Simulation Techniques III, Professional Engineering Publishing, 175-189, 2004.*

H. R. Hamidzadeh, D. J. Nunez, and D. Chandler. Damped response of visco- elastic thick cylinders of infinite extent. *The Proceedings of the Damping 91 Conference on Passive Vibration Darnting and it's Interfaces with Other Disciplines, 3(HBB):, 1-10, 1991.*

H. R. Hamidzadeh and N. N. Sawaya. Free Vibration of Thick Multi-layer Cylinders. *Journal of Shock and Vibrations, 2 (5), 393-401, 1995.*

T. D. Hawkes and K. P. Soldatos. Three dimensional axisymmetric vibrations of orthopedic and cross-ply laminated hollow cylinders. *AIAA Journal 30 (4): 1089-1098, 1992.*

K. H. Huang and S. B. Dong. Propagation waves and edge vibrations in anisotropic composite cylinders. *Journal Sound and Vibration, 96: 363-379, 1984.*

J. R. Hutchinson, Axisymmetric vibrations of a free finite length rod. *Journal of the Acoustical Society of America 51: 233-240, 1972.*

J. R. Hutchinson. Axisymmetric flexural vibrations of thick free circular plate. *ASME Journal of Applied Mechanics, 46: 139-144, 1979.*

J. R. Hutchinson. Vibrations of solid cylinders. *ASME Journal of Applied Mechanics 47: 901-907, 1980.*

J. R. Hutchinson and S. A. El. Azhari Vibration of free hollow circular cylinders. *Journal of Applied Mechanics 53: 641-656, 1986.*

M. K. Jain, S. R. K. Iyengar, and R. K. Jain. *Numerical methods for scientific and engineering computation.* John Wiley & Sons, 1984.

I. W. Jones, and V. L. Salerno. The vibration of an internally damped sandwich plate radiating into a fluid medium. *Journal of Engineering for Industry, 87: 653-662, 1965.*

I. W. Jones, and V. L. Salerno. The effect of structural damping on the forced vibrations of cylindrical sandwich shells. *Journal of Engineering for*

Industry, 88: 318-324, 1966.

B. G. Korenev. *Bessel functions and their applications.* Routledge Publisher, 2002.

H. Lamb. On the propagation of tremors over the surface of an elastic solid. *Philosophical Transactions of the Royal Society of Landon, series A. 203:1-42, 1904.*

B. J. Lazan. *Damping of materials and members in structural mechanics.* Pergamon Press Ltd., 1968.

A. W. Leissa. Vibration of Shells. *NASA SP - 288, 1973.*

R. B. Lindsay. *Lord Rayleigh: the man and his work.* Permagon Press, Oxford, 1970.

A. E. H. Love. *Mathematical theory of elasticity.* 4th Ed, Cambridge University Press,

Cambridge, 1927.

Y. P. Lu, B. E. Douglas and E. V. Thomas. Mechanical impedance of damped three-layered sandwich rings. *AIAA Journal 11(3): 300-304, 1973.*

Y. P. Lu and C. C. Everstine. More on finite element modeling of damped composite systems. *Journal of Sound and Vibration 69(2):199-205, 1980.*

C. P. Lusher and W. N. Hardy. Axisymmetric free vibrations of a transversely isotropic finite cylindrical rod. *Transactions of the ASME, Journal of Applied Mechanics, December 1988, Vol.55/855.*

A. K. Mal, and S. J. Singh. *Deformation of elastic solids.* Prentice Hall,Inc. 1991.

S. Markus. *The mechanics of vibrations of cylindrical shells.* Elsevier, New York, 1988.

G. W. McMahon. Experimental study of the vibrations of solid, isotropic elastic cylinders. *Journal of the Acoustical Society of America 36: 85-92, 1964.*

H. D. McNiven and D. C. Perry. Axially symmetric waves in finite elastic rods. *Journal of the Acoustical Society of America 34: 433-437, 1962.*

H. D. McNiven, A. H. Shah, and J. L. Sackinan. Axially symmetric waves in hollow, elastic rods part I. *Journal of the Acoustical Society of America 40:784-791, 1966.*

D. J. Mead and S. Markus. Loss factor and resonant frequencies of encastre damping sandwich beams. *Journal of Sound Vibration, 12(1), 99-112, 1970.*

B. C. Nakra and P. Grootenhuis. Structural damping using a four layered sandwich. *Journal of Engineering for Industry 81-85, 1972.*

B. C. Nakra. Vibration control with visco-elastic materials. *Schok Vibration Digest, 13: 17-20, 1981.*

V. S. Nakra and P. Grootenhuis. Extensional effects in constrained viscoelastjc layer damping. *The Aeronautical Quarterly 25: 225-231, 1974.*

H. H. Pan. Axisyminetrical Vibrations of a Circular Sandwich Shell with a Viscoelastic Core Layer. *Journal of Sound and Vibration, 9 (2): 338-348, 1969.*

S. T. Pang, T. C. Tsao, and L. A. Bergman. Interaction effects in the hybrid control of Euler-Bernoulli beams. *The Proceedings of 9th ASCE Engineering Mechanics Conf, 820-823, 1992.*

G. G. Parfitt and D. Lambeth. The Damping of Structural Vibrations. *ARC/CP-596, 1960.*

G. Pickett. Flexural vibrations of unrestrained cylinders and disks. *Journal of Applied Physics, 16: 820-831, 1945.*

W. Pietraszkiewicz and, C. Szymczak. *Shell Structures: Theory and Applications.* Taylor and Francis, 2006

L. Pochhammer. Uber die fortpflanzungsgeschwifldigkeiten klejner schwirigungen in elnem unbegrenzten isotropen kreiscylinder. *Z. Math 81:324-336, 1876.*

G. H. Qing, Y. H. Liu, J. J. Qiu, X. J. Meng. A semi-analytical method for the free vibration analysis of thick double-shell systems. *International Journal of Finite Elements in Analysis and Design , 42: 837-845, 2006.*

M. S. Qatu. Theory and vibration analysis of laminated barrel thin shells. *Journal of Vibration and Control, 5(6) 851-889, 1999.*

S. S. Rao. *Mechanical vibrations.* 2Ed Addison Wesley, 1990.

S. N. Rashband. Resonant vibrations of free cylinders and disks. *Journal of the Acoustical Society of America 57 (4): 899905, 1975.*

N. Rattanawangcharoen and A. H. Shah. Vibration of laminated isotropic cylinders. *The Proceedings of Canadian Congress of Applied Mechanics, 1: 230-231, 1991.*

N. Rattanawangcharoen, A. H. Shah, and S. K. Datta. Wave Propagation in Laminated Composite Circular Cylinders. *Int. J. Solids Structures, 29 (6): 767-781, 1992.*

L. Rayleigh. *Theory of sound. volume 1, 2 Ed.* Macmillan Co. London, 1894.

M. Rumerman and S. Raynor. Natural frequencies of finite circular cylinders in axially symmetric longitudinal vibration. *Journal of Sound and Vibration 15: 529-543, 1971.*

R. K. Singal and K. Williams. A theoretical and experimental study of vibrations of thick circular cylindrical shells and rings. *ASME Journal of Vibrations, Acoustics. Stress and Reliability in Design 110: 533-537, 1988.*

J. C. Snowdon. *Vibration and shock in damped mechanical systems.* John Wiley and Sons, 1968.

N. G. Stephen and M. Z. Wang. Decay rates for the hollow circular cylinder. *Journal of Applied Mechanics 59: 747-753, 1992.*

P. A. Svardah. Three-dimensional analysis of axisymmetric transient waves in hollow elastic cylinders. *ASME Journal of Applied Mechanics 51: 792-796, 1984.*

S. P. Timoshenko. *History of strength of materials*, McGraw Hill, New York, 1953.

J. R. Vinson and R. L. Sierakowski. *Behavior of structures composed of composite materials.* Kluwer Academic Publisher, 2004.

G. B. Warburton. Vibration of thin cylindrical shells. *Journal of Mechanical Engineering Science 7: 399-407, 1965.*

G. N. Watson. *Theory of Bessel functions. 2 Ed.* Cambridge, 1962.

J. C. White. The flexural vibrations of thin laminated cylinders. *ASME Journal of Engineering Industry, 397-402, 1961.*

F. B. Yeh and C. D. Yang . New time domain identification technique. *AIAA Journal of Guidance and Control, 10(3): 313-316, 1987.*

Y. Y. Yu. Vibrations of elastic sandwich cylindrical shells. *Journal of Applied Mechanics. 27: 653-662, 1960.*

J. Zhou and B. Yang. A distributed transfer function method for analysis of cylindrical shells. *AIAA Journal, 33(9): 1698-1708, 1995.*

J. Zhou and B. Yang. Three-dimensional stress analysis of thick laminated composite cylindrical shells and panels. *AIAA Journal, 34(9): 1960-1964, 1996.*

Appendix A

Trigonometric Formulae

Definitions in terms of exponentials.

$$\cos z = \frac{e^{iz} + e^{-iz}}{2} \tag{A.1}$$

$$\sin z = \frac{e^{iz} - e^{-iz}}{2i} \tag{A.2}$$

$$\tan z = \frac{e^{iz} - e^{-iz}}{i\left(e^{iz} + e^{-iz}\right)} \tag{A.3}$$

$$e^{iz} = \cos z + i \sin z \tag{A.4}$$

$$e^{-iz} = \cos z - i \sin z \tag{A.5}$$

Angle sum and difference.

$$\sin(\alpha \pm \beta) = \sin\alpha\cos\beta \pm \cos\alpha\sin\beta \tag{A.6}$$

$$\cos(\alpha \pm \beta) = \cos\alpha\cos\beta \mp \sin\alpha\sin\beta \tag{A.7}$$

$$\tan(\alpha \pm \beta) = \frac{\tan\alpha \pm \tan\beta}{1 \mp \tan\alpha\tan\beta} \tag{A.8}$$

$$\cot(\alpha \pm \beta) = \frac{\cot\alpha\cot\beta \mp 1}{\cot\beta \pm \cot\alpha} \tag{A.9}$$

Symmetry.

$$\sin(-\alpha) = -\sin\alpha \tag{A.10}$$

$$\cos(-\alpha) = \cos\alpha \tag{A.11}$$

$$\tan(-\alpha) = -\tan\alpha \tag{A.12}$$

Multiple angle.

$$\sin(2\alpha) = 2\sin\alpha\cos\alpha = \frac{2\tan\alpha}{1 + \tan^2\alpha} \tag{A.13}$$

$$\cos(2\alpha) = 2\cos^2\alpha - 1 = 1 - 2\sin^2\alpha = \cos^2\alpha - \sin^2\alpha \tag{A.14}$$

$$\tan(2\alpha) = \frac{2\tan\alpha}{1-\tan^2\alpha} \tag{A.15}$$

$$\cot(2\alpha) = \frac{\cot^2\alpha - 1}{2\cot\alpha} \tag{A.16}$$

$$\sin(3\alpha) = -4\sin^3\alpha + 3\sin\alpha \tag{A.17}$$

$$\cos(3\alpha) = 4\cos^3\alpha - 3\cos\alpha \tag{A.18}$$

$$\tan(3\alpha) = \frac{-\tan^3\alpha + 3\tan\alpha}{-3\tan^2\alpha + 1} \tag{A.19}$$

$$\sin(4\alpha) = -8\sin^3\alpha\cos\alpha + 4\sin\alpha\cos\alpha \tag{A.20}$$

$$\cos(4\alpha) = 8\cos^4\alpha - 8\cos^2\alpha + 1 \tag{A.21}$$

$$\tan(4\alpha) = \frac{-4\tan^3\alpha + 4\tan\alpha}{\tan^4\alpha - 6\tan^2\alpha + 1} \tag{A.22}$$

$$\sin(5\alpha) = 16\sin^5\alpha - 20\sin^3\alpha + 5\sin\alpha \tag{A.23}$$

$$\cos(5\alpha) = 16\cos^5\alpha - 20\cos^3\alpha + 5\cos\alpha \tag{A.24}$$

$$\sin(n\alpha) = 2\sin((n-1)\alpha)\cos\alpha - \sin((n-2)\alpha) \tag{A.25}$$

$$\cos(n\alpha) = 2\cos((n-1)\alpha)\cos\alpha - \cos((n-2)\alpha) \tag{A.26}$$

$$\tan(n\alpha) = \frac{\tan((n-1)\alpha) + \tan\alpha}{1 - \tan((n-1)\alpha)\tan\alpha} \tag{A.27}$$

Half angle.

$$\cos\left(\frac{\alpha}{2}\right) = \pm\sqrt{\frac{1+\cos\alpha}{2}} \tag{A.28}$$

$$\sin\left(\frac{\alpha}{2}\right) = \pm\sqrt{\frac{1-\cos\alpha}{2}} \tag{A.29}$$

$$\tan\left(\frac{\alpha}{2}\right) = \frac{1-\cos\alpha}{\sin\alpha} = \frac{\sin\alpha}{1+\cos\alpha} = \pm\sqrt{\frac{1-\cos\alpha}{1+\cos\alpha}} \tag{A.30}$$

$$\sin\alpha = \frac{2\tan\frac{\alpha}{2}}{1+\tan^2\frac{\alpha}{2}} \tag{A.31}$$

$$\cos\alpha = \frac{1-\tan^2\frac{\alpha}{2}}{1+\tan^2\frac{\alpha}{2}} \tag{A.32}$$

Powers of functions.

$$\sin\alpha\cos\alpha = \frac{1}{2}\sin(2\alpha) \tag{A.33}$$

$$\sin^2\alpha = \frac{1}{2}(1-\cos(2\alpha)) \tag{A.34}$$

$$\cos^2\alpha = \frac{1}{2}(1+\cos(2\alpha)) \tag{A.35}$$

$$\tan^2\alpha = \frac{1-\cos(2\alpha)}{1+\cos(2\alpha)} \tag{A.36}$$

$$\sin^3\alpha = \frac{1}{4}(3\sin(\alpha)-\sin(3\alpha)) \tag{A.37}$$

$$\sin^2\alpha\cos\alpha = \frac{1}{4}(\cos\alpha-3\cos(3\alpha)) \tag{A.38}$$

$$\sin\alpha\cos^2\alpha = \frac{1}{4}(\sin\alpha+\sin(3\alpha)) \tag{A.39}$$

$$\cos^3\alpha = \frac{1}{4}(\cos(3\alpha)+3\cos\alpha)) \tag{A.40}$$

$$\sin^4\alpha = \frac{1}{8}(3-4\cos(2\alpha)+\cos(4\alpha)) \tag{A.41}$$

$$\sin^3\alpha\cos\alpha = \frac{1}{8}(2\sin(2\alpha)-\sin(4\alpha)) \tag{A.42}$$

$$\sin^2\alpha\cos^2\alpha = \frac{1}{8}(1-\cos(4\alpha)) \tag{A.43}$$

$$\sin\alpha\cos^3\alpha = \frac{1}{8}(2\sin(2\alpha)+\sin(4\alpha)) \tag{A.44}$$

$$\cos^4\alpha = \frac{1}{8}(3+4\cos(2\alpha)+\cos(4\alpha)) \tag{A.45}$$

$$\sin^5\alpha = \frac{1}{16}(10\sin\alpha-5\sin(3\alpha)+\sin(5\alpha)) \tag{A.46}$$

$$\sin^4\alpha\cos\alpha = \frac{1}{16}(2\cos\alpha-3\cos(3\alpha)+\cos(5\alpha)) \tag{A.47}$$

$$\sin^3\alpha\cos^2\alpha = \frac{1}{16}(2\sin\alpha+\sin(3\alpha)-\sin(5\alpha)) \tag{A.48}$$

$$\sin^2\alpha\cos^3\alpha = \frac{1}{16}(2\cos\alpha-3\cos(3\alpha)-5\cos(5\alpha)) \tag{A.49}$$

$$\sin\alpha\cos^4\alpha = \frac{1}{16}(2\sin\alpha+3\sin(3\alpha)+\sin(5\alpha)) \tag{A.50}$$

$$\cos^5\alpha = \frac{1}{16}(10\cos\alpha+5\cos(3\alpha)+\cos(5\alpha)) \tag{A.51}$$

Products of sin and cos.

$$\cos\alpha\cos\beta = \frac{1}{2}\cos(\alpha-\beta)+\frac{1}{2}\cos(\alpha+\beta) \qquad \text{(A.52)}$$

$$\sin\alpha\sin\beta = \frac{1}{2}\cos(\alpha-\beta)-\frac{1}{2}\cos(\alpha+\beta) \qquad \text{(A.53)}$$

$$\sin\alpha\cos\beta = \frac{1}{2}\sin(\alpha-\beta)+\frac{1}{2}\sin(\alpha+\beta) \qquad \text{(A.54)}$$

$$\cos\alpha\sin\beta = \frac{1}{2}\sin(\alpha+\beta)-\frac{1}{2}\sin(\alpha-\beta) \qquad \text{(A.55)}$$

$$\sin(\alpha+\beta)\sin(\alpha-\beta) = \cos^2\beta-\cos^2\alpha = \sin^2\alpha-\sin^2\beta \qquad \text{(A.56)}$$

$$\cos(\alpha+\beta)\cos(\alpha-\beta) = \cos^2\beta+\sin^2\alpha \qquad \text{(A.57)}$$

Sum of functions.

$$\sin\alpha\pm\sin\beta = 2\sin\frac{\alpha\pm\beta}{2}\cos\frac{\alpha\pm\beta}{2} \qquad \text{(A.58)}$$

$$\cos\alpha+\cos\beta = 2\cos\frac{\alpha+\beta}{2}\cos\frac{\alpha-\beta}{2} \qquad \text{(A.59)}$$

$$\cos\alpha-\cos\beta = -2\sin\frac{\alpha+\beta}{2}\sin\frac{\alpha-\beta}{2} \qquad \text{(A.60)}$$

$$\tan\alpha\pm\tan\beta = \frac{\sin(\alpha\pm\beta)}{\cos\alpha\cos\beta} \qquad \text{(A.61)}$$

$$\cot\alpha\pm\cot\beta = \frac{\sin(\beta\pm\alpha)}{\sin\alpha\sin\beta} \qquad \text{(A.62)}$$

$$\frac{\sin\alpha+\sin\beta}{\sin\alpha-\sin\beta} = \frac{\tan\frac{\alpha+\beta}{2}}{\tan\frac{\alpha-+\beta}{2}} \qquad \text{(A.63)}$$

$$\frac{\sin\alpha+\sin\beta}{\cos\alpha-\cos\beta} = \cot\frac{-\alpha+\beta}{2} \qquad \text{(A.64)}$$

$$\frac{\sin\alpha+\sin\beta}{\cos\alpha+\cos\beta} = \tan\frac{\alpha+\beta}{2} \qquad \text{(A.65)}$$

$$\frac{\sin\alpha-\sin\beta}{\cos\alpha+\cos\beta} = \tan\frac{\alpha-\beta}{2} \qquad \text{(A.66)}$$

Trigonometric relations.

$$\sin^2\alpha-\sin^2\beta = \sin(\alpha+\beta)\sin(\alpha-\beta) \qquad \text{(A.67)}$$

$$\cos^2\alpha-\cos^2\beta = -\sin(\alpha+\beta)\sin(\alpha-\beta) \qquad \text{(A.68)}$$

Index

Lightning Source UK Ltd.
Milton Keynes UK
14 March 2010

151358UK00007B/164/P

9 780387 755908